Total Relativity
And
Dimensional Dynamics

**A different way to think about the universe:
space, gravity, light, and time.**

Evert C. Schoevers

Copyright © 2024 (Evert C. Schoevers)
All rights reserved worldwide.

No part of the book may be copied or changed in any format, sold, or used in a way other than what is outlined in this book, under any circumstances, without the prior written permission of the publisher.

Publisher: Inspiring Publishers,
P.O. Box 159, Calwell, ACT Australia 2905
Email: publishaspg@gmail.com
http://www.inspiringpublishers.com

 A catalogue record for this book is available from the National Library of Australia

National Library of Australia The Prepublication Data Service

Author: Evert C. Schoevers
Title: Total Relativity and Dimensional Dynamics
Genre: Non-fiction

Paperback ISBN: 978-1-922920-73-7
ePub2 ISBN: 978-1-922920-74-4
PDF eBook ISBN: 978-1-922920-75-1

Contents

What is TRADD? 1
   The first 40 years 2

What is space? 8
   Dimensional dynamics 8
      Two-dimensional Space 9
         Conserving the boundary 9
         Conserving the dimension 11
         Deformation of two-dimensional Space 12
      Three-dimensional Space 15
         Conserving the boundary 15
         Conserving the dimension 16
         Deformation of three-dimensional space 17
      Four-dimensional Space 18
         Innovating the hypercube 19
         Deformation of four-dimensional space (varying the three-dimensional boundary) 24
   Interim conclusion 29
   Expanding space 30
      How might space expand? 30
         Expansion in the x-w plane? 33
            Moving a sheet in the x-y plane in the z-direction 34
            Moving an x-y plane in the z-direction 37
            Moving an x-y plane in the w-direction 38
            The x-w plane 40
         Introducing forces 41

  Relative strengths 44
 Introducing time 46
  Creating time 46
   Trick or treat? 47
   Travelling through time 48
   Measuring time and distance 48
   Renaming the w-axis 49

What is gravity? 51
The standard model 51
 Bent straight lines 53
Newtonian gravity 53
Getting rid of gravity 54
Reinventing gravity 55
 How Einstein killed gravity 55
 Gravity: matter and space 58
  Einstein's map of space and time 58
  Holes in space 60
  Space that flows 61
 Bending space without gravity 62
  Action at a distance 63
 There are no holes! 65
  Expanding space 65
  Contracting space 65
  When space contracts 66
  Leaking space without holes 66
 Spacetime is curved … apparently 67
 Mass in space 68
 Introducing a point mass 71
  Lengths contract 72
  Time slows 73
  Proximity smoothing 75
  First time, then length 76
 Introducing a point mass again 78
  Mass slows time 79

    More mass slows time more 80
    Contracting space contracts lengths 81
    Mass changes time changes space changes length 84
  Mass mediates force 85
    Mediated force slows time 87
    Mediated force contracts space 88
    Contracting space is gravity 88
    Comparing TRADD with the standard model 88
  Mass – falling through spacetime. 90
    Reintroducing a point mass 90
    How time slows down 92
    Falling through time 93
    Falling through space 94
    Falling through spacetime 95
    Reconstructing the criterion 96
  Flattening time 97
    When space is bent in the time direction 97
    Straightening out the curves 98
  So, is gravity a force or not? 101

What is light? 104
  The speed of light 104
  A light-year 104
  Something doesn't add up 104
  Spacetime 106
  How does light move? 107
    When we don't move 109
    When moving away from light 114
      How we move through spacetime 117
        The faster we go, the faster we go! 118
        Getting there sooner: time-drops 120
        Finding a photon's flight path 121
          Designing an experiment 121
          Photons: speeding them up 122
          Measuring approach speed 123

- Incorporating time-drops 125
- Time-dropped trajectories 126
- A photon's curved flight path 130
- When moving toward light 132
  - Designing an experiment 132
  - Photons: slowing them down 133
  - Measuring approach speed 133
  - Incorporating time-drops 134
  - Time-dropped trajectories 134
  - A photon's curved flight path 135
- Toward, away from, or waiting for a photon 136
- More questions! 137
  - Waiting for a photon 138
  - Photons from less than a light-year away 139
  - Photons from more than a light-year away 140
  - Photon: It's coming – It's here – It's been and gone 141
  - Photons that keep travelling 142
- Faster through space than time 147
- Light is energy that travels as a wave 148
- The whole picture 149

What is time? 151

- The characteristics of time 152
  - Time is a direction 152
  - We perceive time because we move through it. 153
  - We travel through time at light-speed 153
  - When we travel through time, space expands 154
  - Expanding space manifests change 155
  - Change is consistent with the 'flow' of time 156
  - The mechanics of spacetime 157
- Accommodating time 158
  - Velocity 159
  - Our velocity through spacetime 160
    - When our velocity through space is zero 163
    - When our velocity through space is 0.25 light-speed 164

    When our velocity through space is 0.5 light-speed 165

    When our velocity through space is 0.75 light-speed 166

    Our speed through spacetime is constant 167

    We travel at the speed of space 168

    The why and how of time dilation 168

    Calculating time dilation in the TRADD model 169

Final thoughts 171

Notes 174

Acknowledgments 175

Introduction

What is space?

Science never gave us an answer, then avoided the question. Einstein did away with space when he postulated Minkowski's spacetime continuum. Yet the space in spacetime contracts when we move (Special Relativity), and contracts and bends due to mass (General Relativity). If space can do that, then perhaps it is "something" rather than "nothing". TRADD restores space as "something", then asks the question. The answer will surprise you.

What is gravity?

Newton calculated gravity as a force between matter and matter. Einstein crafted gravity from a force between matter and space. In TRADD, gravity emerges from forces between space and time.

What is light?

Particle? Wave? Or something else?

There is something special about the speed of light; 299,792,458 metres per second, that may have nothing to do with light and everything to do with metres and seconds, (i.e., with space and time). Is the constant speed, 299,792,458 metres per second, a property of light, or is it a fundamental property of space and time themselves? It's just a ratio, right?

Einstein left us a clue in his equation $E=mc^2$. In it, the speed of light is associated with energy and mass. And what do energy and mass do? They bend spacetime. Again, we are left to wonder if this number - the speed of light - is not a property of energy or mass, but of space and time.

What is time?

Again, no clear answer from science. Einstein got rid of time too. But even with time bound up in the spacetime continuum, scientists continue to ask what time is. Time in spacetime slows when we move (Special Relativity) and slows due to mass (General Relativity). Time seems to flow, but only in one direction. TRADD finds time hiding in plain sight, a cog in the mechanics of the universe. Astonishing!

What is 'Total Relativity And Dimensional Dynamics', (TRADD)?

Coming from a fresh perspective, this book will guide you to think about the universe in a different way. What space is, how gravity works, why the speed of light has the value that it does, and if we can truly understand time.

What is TRADD?

TRADD stands for 'total relativity and dimensional dynamics'.

It is a fanciful title I came up with in my late twenties or early thirties – I forget which. You know? You set off thinking about space and time. You write notes. You make drawings. You start filling up a folder full of notes and drawings. Then you feel the folder needs a title. I came up with TRADD. I've lived with it for 40 years and it has become synonymous with the contents of this work. Even though I'd like to find a more suitable title – it's just too much sentiment to let go of – so it stays. The subtitle is more descriptive. I was kindly advised to use that because it tells you what the book is about. When I thought up TRADD as a title for my folder it was full of ideas to do with the dimensions of space and questions about gravity and the universe expanding - the realm of general relativity. I was thinking about space but had to contend with the idea that general relativity cast our universe as a spacetime continuum. It was the *time* component, that for me, didn't belong – indeed that's where it all began.

TRADD is a collection of ideas that loosely comprise a cosmological model that's a bit different to what you might expect in some ways and exactly what you'd expect otherwise.

So, what should you expect? Well, not a rigorous scientific theory with lots of mathematics, equations, predictions, and proclamations. But don't expect it to be an easy read, either – some of it is difficult – sorry, but some things in their simplest form demand something extra from us. Instead, expect to share in what insights emerge when we think carefully about what we already know, but in an unconventional way, then – expect to be fascinated.

Science works best with 'what is' and with 'what can be proved'. (And it does this by demanding an idea be falsifiable. In other words, if you are unable to prove something wrong or not, then proving it *right* has no currency.) Science does best with what happens and what can be predicted to happen – cause and effect – specifically, stuff we can measure and stuff we can observe. *Why* and *how* something happens is often elusive, more often evades proof, and almost always escapes measurement. Hence, science falters when asked to deal with 'why' and 'how'. *Why* does the universe expand and *how* does it expand? *Why* is the speed of light 'that' speed and *how* does it move so we always get 'that' value when we measure it? *Why* does mass warp spacetime and *how* does it do that? *Why* does time slow down in a gravitational field and when we speed up and *how* does that happen?

Science will mostly point to equations to explain the 'why' – '… because that's what the math tells us'. As to the 'how' things happen, science offers us less than we'd like.

This book is an honest attempt to put forward *suggestions*, as possible answers to questions dealing with *why* and *how* things happen; the *why* and *how* things happen with space, gravity, light, and time. When considering how to integrate these with what science already knows, (the standard model), I was surprised they remained consistent with one another in a broader context. Even as I strived to associate these different ideas, I was nevertheless taken aback with the results. I didn't expect different ideas about *why* and *how* space expands to be consistent with different ideas about *why* and *how* we have gravity. Different ideas about *why* and *how* light is measured as having the speed it does were accommodated by the different ideas about space and gravity. And the different ideas about the *why* and *how* of time were also accommodated. Not only were these different ideas consistent with one another, they seemed to be consistent with the standard model as well – with what we already know. A cosmological model framing these different ideas as *suggestions* subsequently emerged. The model isn't complete. Questions remain. But if you run out of road, that doesn't have to mean you're lost. It might just mean that no one's gone that way before.

The *suggestions* put forward here paint a somewhat cohesive picture of *how* our universe might be constructed and *why* it might operate as it does – its *mechanics*.

The pursuit of TRADD began with a rejection of the idea that time was 'the fourth dimension'. What did that mean? Were they saying that time was a *spatial* dimension?

TRADD was developed over many years. Coming up with ideas, testing them, and finding they were 'not true'. And then coming up with new ideas, testing them, and finding they were not 'true'. For an idea to be true, it had to fit with known science, or at least explain how it deviated from what was known in a legitimate way. The initial goal was not to come up with a cosmological model. It was to explain, to myself, why time was not the fourth dimension. And to do that, I had to explain *what* it was. And whatever I came up with had to make sense. I was not satisfied to simply find an alternative for time as the fourth dimension. I wanted what I found to make sense. And for that, it had to make sense to everyone, not just to me. It had to stand up to scrutiny.

Now whether it does or doesn't stand up to scrutiny, I'll leave to others. I'm not a scientist. I'm not even an author. TRADD could be nothing more than a collection of ideas invented to make sense of reality – my reality.

This is the story of how I found time.

The first 40 years

I became aware of relativity theory and the expanding universe as a young boy. At around 10 or 12 years of age, in the 1960s, I thought deeply about the idea of time being a dimension. I thought I understood what a dimension was but could only visualise spatial dimensions. The idea that time was also a dimension was not something I could come to grips with or easily accept. I rejected the idea. Yet, it troubled my mind for years to come.

Our universe, modelled under Einstein's theories of special relativity and general relativity, is described as a spacetime continuum. The model has three spatial dimensions (x, y, and z) and one time dimension (t).

With the time dimension embedded in the model as it was, I knew that to reject it, I would need to remove it. To do that, I would first need to come to grips with the model itself. I would question what *right* time had to be there – in a bundle of spatial dimensions woven into it. Thus, early development centred on the accepted model at that time. What science was saying about what the *expanding* universe was, as well as how, why, and if the universe was expanding. What was it expanding out of? What was it expanding into?

To do that, my efforts were directed at gaining a better understanding of what space was. If the universe was described as spacetime, and if I didn't like the idea that time was part of that description, I would need to come up with a description that did not involve time. That was easy when thinking in one, two, and three dimensions. We all do it. But adding a fourth dimension increased the level of difficulty – a lot. After that, having constructed a model of *four*-dimensional *space*, entertaining the idea that space might be expanding presented new challenges.

Construction of a model in four-dimensional space had been done before. It had been done geometrically, mathematically, and philosophically. I was not the first. But my approach was different. Indeed, I constructed my own drawing of a hypercube years before I saw one. And I must tell you – mine looked nothing like the one in the book. I had to label the end points of each line in both drawings (mine and the one in the book) to see they were the same. Mine was simply less pleasing to the eye and most of my lines were curved, not straight. I eventually ended up with a model of four-dimensional space that could expand and contract. It could be poked and prodded to refashion its expanse and its boundaries. It could literally shape-shift. But there was a problem. There were rules.

Moving the model from something playful on paper to something that described space in our universe should have been straightforward; or so I thought. Not only did the rules of the paper model persist, but they also, now, had to *mean* something. Stretching the length of a square to make a rectangle is easy on paper. You just use more paper. But to do that in the universe, you can't simply say, 'Oh? I'll use more universe'. That wouldn't do. On paper, it was me that stretched the length of the square. First, I decided to do it. Then it was *me* that did it. Easy – square, rectangle. But in the universe, how is the decision made to stretch the length of a square? What does the stretching? And how is it that using 'more universe' is something that is allowed – if it's allowed?

I can tell you: I was getting very frustrated. I had created a model in four spatial dimensions. Fantastic— no mention of time. But only to find that the very rules I had invented to make the model work could not simply be thrown away when I wanted to use the model to describe space in our universe. No, the rules had to stay. And the rules had to mean something. And seriously, I had no clue.

Months passed. Then it dawned on me. Matter! The universe was not *just* space. It contained matter. Matter had mass and mass was energy. Our universe contained energy. And now the rules of the model had a focus. If the rules had to mean something, then that meaning (I reasoned) had something to do with energy. So, I worked to explain the rules of the universe because they were no longer the rules of the paper model. The paper model didn't have energy. The universe did. Then something stopped me in my tracks. I ran into a massive, mega, ugly, disgusting, should-not-be-allowed brick wall—gravity!

Nothing in the model was able to accommodate this beast. And a beast it was. I had set out to model space without time. I did that with a satisfying measure of success. I had transitioned the model from paper to the real world—an achievement, I thought. Gravity crushed it, brutally, as if the paper model were ripped from the sketch pad and binned. I had heard stories about gravity crushing things all over the cosmos. On that day, I became a believer—I was crushed.

Whilst learning about gravity, I came to understand and appreciate a little more about Einstein's theory of general relativity. But it contained two agitating elements—time and gravity. Specifically, the idea that gravity acted over a distance. I had already rejected time as a dimension; that drove me to find an alternative description of space. Now, the idea that gravity could pull on distant objects with a force—that daunted me. You need a rope, right? Something! Sure, I later learnt about the difference between Newtonian gravity and Einsteinian gravity; from objects pulling on each other to objects warping spacetime and moving through it accordingly. But I was not convinced. There was still the matter of distance. And whilst I felt I understood it and I knew that predictions were tested and found to be true, I felt uneasy with it. As if something was *not quite right*.

It was around 1984 when I began modelling in four dimensions. My goal then was to create a universe without time. Well, at least that the description of the space in the universe did not include time. By around 1994, I found myself needing to find a second description. I needed to describe gravity—in a different way. I found myself in a position of not *only* needing to account for space in a different way—without time, but gravity in a different way—not acting over a distance. The new description of gravity took about 10 years.

The model was initially a new description of space. A new description of gravity was added to it. One that fitted the model and obeyed the rules used to create it. The model was becoming a new way of describing the universe, not only space. By around 2004, the speed of light presented yet another hurdle. Significantly, that the approach speed of light as measured by an observer was the same, regardless of the observer's motion. It didn't fit the model. I needed a third description. I needed to describe light in a different way. It had to fit into the new model of space and gravity but retain its key attribute— 'constant' speed.

I had developed a new description of gravity and how its effects could be interpreted by science the way they were. That was important to me. I never set out to prove the existing science wrong. Indeed, I was more than satisfied that science had got it right. I just wanted to find an interpretation that made more sense—to me. The alternate description of gravity I developed would not change what we measured but would provide a different explanation about what was happening: the *why* and the *how* of it. But the new description of gravity (yet not a clue about what time was) would not accommodate the idea that the speed of light was constant. If light moved through this new description of space with its new description of gravity, its *measured* speed would need to remain constant—but in a way that could be explained—with an explanation that made sense.

I cannot understate how mesmerising this task became. The new description of light would need to show *why* its speed is constant. I reasoned that if light travelled at a constant speed—in the way that we measure it—then the speed itself *must* be significant. So, the new description would also need to explain why light travelled at *that* speed. Then, in rare brief moments when I thought the puzzle solved, ever-present distractions emerged. Light could be measured as a wave of energy. It could also be measured as a fast-moving particle—a photon. Light, became for me, the most elusive member of the cosmos. It masquerades as a wave, as a particle, and then, as both. It travels at a speed that cannot be surpassed, and always at the same speed, however it's measured. Smoke and mirrors—magic. Simple as sunshine, yet incomprehensible.

We see the universe because light travels to us, through it. General relativity theory describes how mass bends the space through which light travels. We then see the path of light through space appear to bend as it travels through bent space. It was around 2014 when I began to look deeper into what light was and how we understood it. I became intrigued with the dual particle-wave theory—the idea that light could be thought of as a wave but also as a particle. I was fascinated to learn that quantum electro dynamics (QED)[1] insists light is a particle—a photon. Then, described as a wave, I was bewildered at the idea of that. I began looking at waves. The kind we are all familiar with. Drop a pebble into a pond. The wave spreads out! It propagates in *all* directions. That's quite a bit different from the idea of a particle traveling from point to point, isn't it? And light is polarised. Well, if it's a wave, it is. But how can light be polarised if it's a particle? So many questions.

But try as I might, and did, there was nothing in QED that helped to explain why the speed of light was constant; only that it was. So, the list for my model now had three items. I needed to account for space in a different way—without time, gravity in a different way—so it didn't act at a distance, and light in a different way—so it could travel as a wave or particle at a constant speed through the new description of space with its new description of gravity.

These new descriptions forced my hand as it were, to find new descriptions for other things as well. The new description of space had to account for our observations that the universe was expanding. Expansion of space happens at a *rate* and so time enters as a metric, even as I sought for a new description of it. Any new description of gravity must account for energy and mass because the force gravity represents is related to these. This was interesting to me because I set out believing that the rules about how the new model worked had something to do with energy. And a new description of light must include time because we are not only concerned with its wave and particle characteristics but also its speed. Try calculating speed without time—that's right, you can't. So, it was impossible to describe space, gravity and light, and not touch on matter, mass, energy, and time. I wanted to understand how they could be described and related in a way that made sense.

It came as a surprise to me, then, when I found a description for time, to find that, in a way, it *was* the fourth dimension. But in an incredibly significant way, it also was not. I had finally reached my goal. I could describe space without time—and time in a different way.

I first had to describe space in a different way—without time.

Then describe gravity in a different way—so it would not act at a distance.

Then describe light in a different way—so wave, particle, and constant speed made sense.

Then describe time in a different way—because that was the goal.

I was pleased to end up with different descriptions of space, gravity, light, and time, in a way that all four descriptions co-existed, whilst allowing us to continue to measure them as we do.

The different description of gravity doesn't change the way we measure it. The different description of light doesn't change the way we measure *it*. Describing time in a different way doesn't change our notions of time, nor alter any time-related measurements we currently make. And where space, gravity, light, and time are related in the standard model, those relationships persist.

Initially, to find an alternative explanation for time so that it didn't need to be the fourth dimension in the fabric of spacetime, I set out to build a new model of space that could accommodate a new notion of time. I needed the dimensions of space to be entities having properties that governed their interactions. To do that, I had to discover a new way to view the dimensions of space and account for the force we call gravity. I needed gravity to behave in a way that it did not act on objects at a distance. I needed light to move through space in a way that ensured its approach speed to observers, moving or stationary, was constantly the same value. And I needed to account for *why* light travelled at *that* speed. I needed the way light moved, the way spatial dimensions changed, and the effects measured as attributed to gravity, to be consistent with what the model says the universe is and what the model says about how the universe behaves.

I ended up with a different description for space, then for gravity, then for light, then for time. The result was a cosmological model with changed perspectives on space, gravity, light, and time. I named the model 'total relativity and dimensional dynamics'. 'Relativity' acknowledges the persistence of that aspect of relativity theory that demands objects and events exist and happen *relative* to other objects and events. 'Dimensional dynamics' is my way of describing the interactions between spatial dimensions. 'Total' attempts two things. It invites us to sum relativity theory and dimensional dynamics, but it also emphasizes the relative relations between things as totally pervasive – that the relativistic nature of whatever we observe and measure, as well as what we fail to observe and measure – is absolute. Even time itself is relative. But surprisingly, not in the way you might think. Not quite.

When I refer to the standard model in this book, I am referring to a collection of theories that describe the natural world. General relativity (GR), special relativity (SR), quantum mechanics, quantum electro dynamics (QED), quantum field theory, quantum chromo dynamics, and the standard model of particles. Collectively, these theories enjoy consensus support; most of the scientific community agree that they currently best describe the natural world. They are sometimes referred to more generally as relativity theory and quantum theory.

The TRADD model does not seek to challenge the standard model. It does not seek to diminish it. It is an attempt to show us that the standard model can be looked at in a new way, through a different lens. At the very least, it may help us to think about what is known, in a different way.

Under TRADD, the *apparent* force we call gravity does not act at a distance. Nothing does.

In the following pages, I will describe TRADD in detail. I will share the thought processes and reasoning that lead to a new model of our universe. You will see how it describes the space in our universe in a new way. You will see why a new description of gravity was needed and how it was found. And you will see a new description of light that makes sense of the constant value we get when we measure its speed. Finally, I hope you will appreciate the new description of time - what and why it is - but also that time, light, gravity and space, as described in TRADD, in no way changes the way we measure them. Only the way we understand them – differently.

What is space?

'Space is the dynamic interaction between spatial dimensions.'

Dimensional dynamics

It may take some unlearning and relearning to conceptualize a universe having four *spatial* dimensions. I know from my own experience this has been true. However, it is the only requirement needed to explore what such a universe would be like, how it would behave, and how that behaviour compares with what we observe and believe we know.

What we are going to do here, will, in the first instance, seem quite objectionable. Rather than considering the space around us as, well, just space – we are going to endow it with attributes beyond those we naturally confer. Normally when we think about space – or spaces – we do so in a context where space exists in relation to objects. The space between two buildings. The space in a room. The space around us. When we measure space, we think about length, area, and volume. We think about size, and assign units of measure: metres, square metres, and cubic metres. And we think about shapes.

But consider this.

We never really think about space as being '*something*'. A house is something. A bus is something. The moon is something.

But space?

It's '*nothing*' – right?

General relativity tells us that space and time together form a spacetime continuum. It tells us the *shape* of spacetime changes in the presence of mass. Think about what that means. Space warps according to its proximity to mass; it gets bent out of shape.

Strictly speaking, it is spacetime that undergoes this warping. But when we look out into the universe and say we are looking out into space – well it is *that* space (together with time) that warps.

Now, can you still think about space as being '*nothing*'? Do you really believe that '*nothing*' can get bent out of shape? Or does space now need to be '*something*'? Because '*something*' *can* get bent out of shape.

Our starting point is to consider that space is not '*nothing*' but '*something*'.

Again, strictly speaking, we are saying that spacetime is 'something'. But we are only interested in the space component of spacetime here. We want to explore what space would look like if it were '*something*' instead of '*nothing*'. Then look at what that might mean when general relativity tells us it gets bent out of shape.

Science tells us our universe is expanding. So, we know that the size of our universe, as measured in its spatial dimensions, is getting bigger. It is growing. Science reaches that conclusion by measuring the growing distances between galaxies. It does not measure space. It measures distances between objects *in* space. But if we first say that space is '*something*', then we can explore what might be happening to it when it expands.

When we say that space is expanding, another way to think about this is that spatial dimensions are being *generated*. Where space is contracting, we can use the opposite term and think about spatial dimensions *degenerating*. For the sake of a name, I refer to changes in spatial dimensions as 'dimensional dynamics'. We want to explore *what happens* to space when we say it is '*something*'.

Rather than jump right in and explore a four-dimensional space in this way, we can start with something a little more familiar – two-dimensional space.

Two-dimensional Space

To understand what is meant by degeneration, it is easier to begin with a two-dimensional space. Take a square that has four sides, each seven units in length. The area of the square is the product of length and height: 49 square units. The perimeter of the square is the sum of the lengths of its sides: 28 units.

Conserving the boundary

We want to squash the square so that its height is reduced from seven to six units, but we want to maintain a constant perimeter value of 28 units. When we do that, the reduction in height gives way to an increase in length. If the height is now six units, then the length must be eight units. When we calculate the area of the resultant rectangle, the product of six and eight, we get 48. By squashing the square, we have reduced the area from 49 to 48 square units.

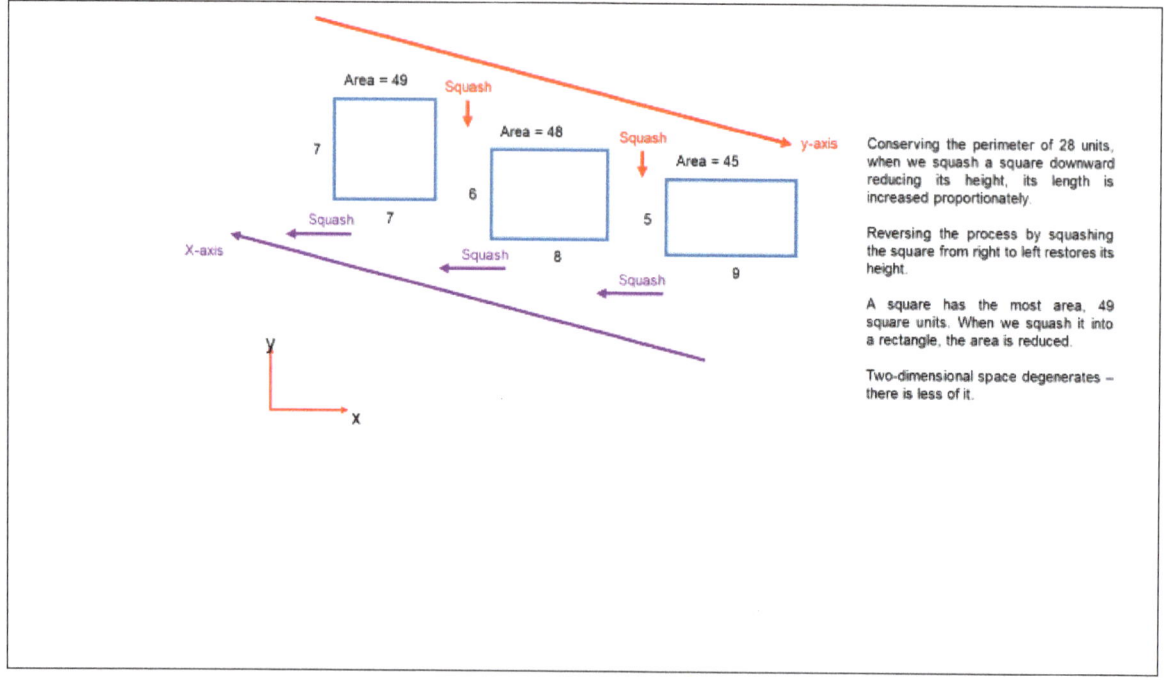

Figure 1.

Area and perimeter are attributes of a two-dimensional space. In our example, squashing the square can be thought of as changing the shape of the two-dimensional space. We changed its area but not its boundary value – its perimeter. But that's because we created a rule.

We said we did not want the perimeter length to change. And so, we maintained it at a length of 28 units. That a two-dimensional space is defined uniquely by its area, allows us to describe a reduction in area as *degeneration* and an increase in area as *generation*. After degeneration, there is less area – there is less two-dimensional space. After generation, there is more area – there is more two-dimensional space. We can see that maintaining a constant length for its boundary – its perimeter – has no conserving effect on the amount of area there is after we squash the square. Reverse the squashing of the square and area is restored from 48 to 49 square units – more area – generating more two-dimensional space.

In this exercise, if we keep squashing down on the square, the height will keep reducing. As the length of the rectangle increases, eventually the height will be zero. At that point, we no longer have a two-dimensional object – we have a line – a one-dimensional object. And yes, the line will be 28 units long. Because that was our rule – to maintain the perimeter at length 28. But having squashed the height to zero, we no longer have a two-dimensional object. So, what is the length of 28 units the perimeter of? Clearly, we have destroyed our two-dimensional object, and so the line is no longer a perimeter – it no longer bounds two-dimensional space. Having destroyed it, we can't simply squash the line from right to left and recreate it. We *could* reverse squash a two-dimensional object and generate more area. But we can't reverse squash a one-dimensional line and generate two-dimensional space.

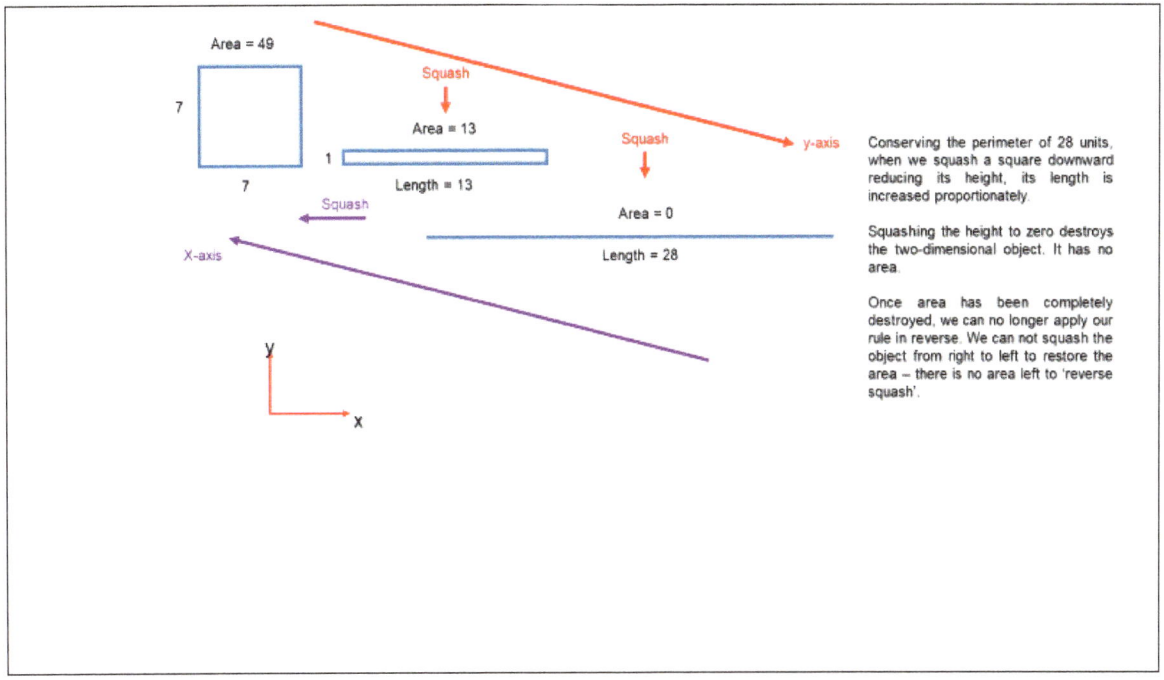

Figure 2.

Conserving the dimension

Well, hang on a minute! Why were we doing this? We were exploring what 'space' *is* by starting with something simple – two-dimensional space. And all we succeeded in doing was to show that if we squashed it and kept squashing it, eventually we would destroy it. But that was because we made up a rule – to conserve the length of the perimeter – the boundary of two-dimensional space. Maybe we made up a wrong rule. Could we make up a different rule instead, so that we can squash our object but never destroy it? Turns out we can. I have already tested the rule that we *don't* conserve the perimeter. That does not get us anywhere – we still end up destroying the object when we squash it. So, let's try a different rule. Let's conserve the area of the object – let's conserve the two-dimensional space when we squash it and see what happens. At least we should expect to not destroy our two-dimensional space because we have made a rule to conserve it.

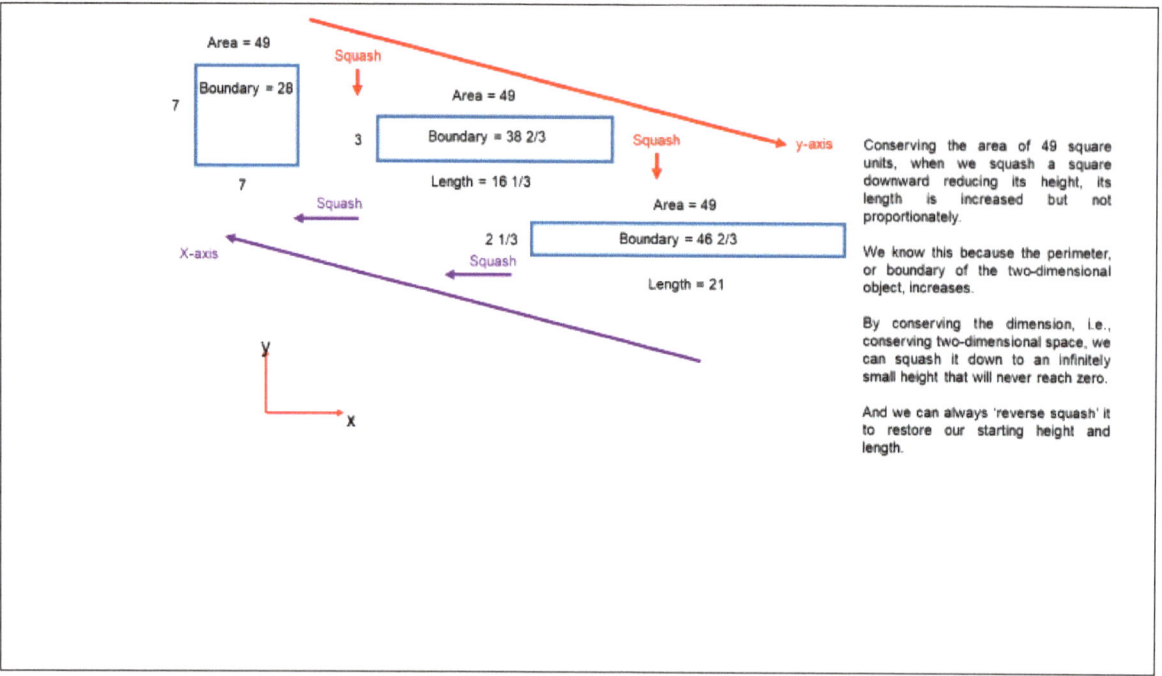

Figure 3.

Our rule works! Squashing the object does not destroy it. We get a longer and longer rectangle, but the height of the object can never reach zero. Our rule forbids it.

In fact, it's all so clever really because we can now squash what was our square, down to a rectangle that has almost infinite length. That's because by conserving our two-dimensional space, our 'area', we can squash the height down, forever getting closer to zero but never reaching it. If we reached it, our area would become zero. So, we are only allowed to make the height smaller and smaller and smaller still. And when we do *that*, conserving our two-dimensional space of 49 square units, the length gets longer and longer and longer. It approaches infinity.

And - bonus points! We *can* 'reverse squash' our object and recover our original height and length. That's because we always have a two-dimensional object to 'reverse squash'. It is never squashed down to a one-dimensional line. With our new rule, if we have a two-dimensional space, squashing it can't destroy it. It doesn't degenerate.

Deformation of two-dimensional Space

When we think about spatial dimensions, we notice that there is always a minimum boundary configuration at which the maximum spatial dimension occurs. For two dimensions, a circle bounds the maximum area for a given perimeter length. Deform the shape of the circle whilst keeping the perimeter length unchanged and the quantity of area is decreased. For three dimensions, the surface area of a sphere bounds the maximum volume. Deform the surface area without altering its size and the volume it bounds is reduced.

Maintaining the boundary of the dimension and then deforming the object's shape results in a degeneration of the spatial dimension.

We are using cubes and squares in place of spheres and circles mainly because the arithmetic is easier to follow. But the principle for these shapes holds, too. Change the dimensions of a perfect square and maintain the length of its perimeter – and keeping the object as a rectangle – and it is found that the area within the object is reduced. Changing the dimensions of a cube and maintaining its surface area also decreases the volume of the resulting object.

In previous operations on an object, maintaining the size of its boundary whilst changing its shape reduced some of the space defined by the object – space was lost. The space bounded by the object vanished and could no longer be accounted for. Returning the object to its former shape increased its amount of bounded space once more.

We need to think carefully about this. On paper, drawing shapes, this is quite legitimate. But we are exploring space as though it has an existence in the real world. On paper, increasing or decreasing the space that an object has is what we need to do so that we can use those drawn objects to describe something. But in the real world, the idea of a reduction in space – well, perhaps that's a thing – but to increase space – can that be a thing, too? Nothing about our real-world experience accords with the idea that something can be produced from nothing. And we also see from our experiments that when something is reduced in quantity, it just doesn't disappear – the reduced quantity never becomes 'nothing'. Consistently, we see that a reduction in a quantity of something results in an increase in some other quantity – a transformation.

If we change the shape of an object, as before but allow the size of its boundary to alter, conserving instead the amount of space bounded, we get a different result worthy of attention.

Take a square that has four sides, each 10 units in length. The area of the square is the product of length and height: 100 square units. The perimeter of the square is the sum of the lengths of its sides: 40 units. Then we squash the square so that its height is reduced from 10 to 4 units, but we want to conserve its area, valued at 100 square units. The reduction in height must give way to an increase in length. We can calculate that the length must be 25 units. By squashing the square, we have increased its boundary from a perimeter of 40 units to 58 units. Area and perimeter are attributes of a two-dimensional space. Squashing the square can be thought of as changing the shape of the two-dimensional space but not its spatial value, its area. That a two-dimensional space is defined uniquely by its area, allows us to associate a reduction in area with degeneration and an increase in area with generation.

But neither occurred. We kept the area constant at 100 square units.

But if we apply the same reasoning to one-dimensional space, we will arrive at the proposition that an increase in length would represent a generation of the dimension and a reduction in length would represent a degeneration of the dimension. By squashing a perfect rectangle – a square – and conserving its area, its one-dimensional boundary, its perimeter, must increase.

If we start with the size of a dimension, say two dimensions of, say, 100 square units – then it is those 100 square units of area that define the dimension. When we shape that area into a square, we end up with a minimum boundary. Its perimeter will have the shortest possible length for that shape and that dimension. That's because a square is a uniform shape – each side has the same length.

The most 'economical' shape is a circle. If we shape our 100 square units into a circle, its boundary (circumference) will have a length shorter than the perimeter of the 100 square unit square.

The square will have a boundary of 10 + 10 + 10 + 10 = 40 units.

To find the boundary of our 100 square unit circle, we first need to calculate the radius.

The formula for the area of a circle is: $A = \pi r^2$. We already know A=100, so we can find r. We simply rearrange the equation so that $A/\pi = r^2$, and then again so that $\pm r = \frac{\pm\sqrt{A/\pi}}{1}$

The formula for the boundary, the circumference is $2\pi r$.

We now have a value for r and so when we multiply r by 2π we get around 35.5.

So, we see that the 35.5-unit boundary of the circle is less than the 40-unit boundary of the square.

Both shapes have the same two-dimensional area of 100 square units. The circle is just a more economical way of bounding the space – it uses a shorter line. When a dimension is arranged into an economical shape, the longest distance between two points in the shape is as short as it can be. We expect, therefore, that the diameter of the circle is shorter than the diagonal line between opposite corners of the square. Let's see if that's true.

Using the equation $a^2 + b^2 = c^2$ with our square where 'a' is the length, 'b' is the height and 'c' is the diagonal length along the line between the opposite corners we want to find, we can do the calculations to get a value for the diagonal of around 14.14.

For our circle, the diameter is simply twice the radius. We can divide our $2\pi r = 35.5$ circumference by π to get $2r = 11.28$, if we leave off a bunch of decimal places.

So, we did it! We were able to show that 11.28 is shorter than 14.14; the longest distance between two points in a circle is less than the longest distance between two points in a square.

We can now introduce a generic term – *deformation*.

When we change the shape of something, we deform it. When we squash our square – we can say we are deforming it. And if we squash a circle, much like squashing a balloon, we will say we are deforming it.

When we think about the shape of a spatial dimension, we like to think that it exists ideally in its most economical form. So, a point for zero dimension, a line for one dimension, a circle for two dimensions and a sphere for three dimensions. But as we have seen, the numbers are easier to visualise and work with if we use lines, squares, and cubes.

When we deform a spatial dimension, so long as we have a rule that its dimension must be conserved, and so long as we start with an economical shape, we end up increasing the size of its boundary. Its boundary is always the next lower dimension. So that deforming a spatial dimension increases the magnitude of the next lower dimension – the deformed spatial dimension's boundary.

Three-dimensional Space

We now want to do with three-dimensional space what we did with two-dimensional space. We want to squash it. We can use a cube as an example of a three-dimensional space. Our cube has edges that are seven units long. Before we squash our cube, we note its boundary is a measure in the one lower dimension, two-dimensional area, commonly called surface area. The cube has six faces, each having an area of 49 square units, which sum to 294 square units. So, our three-dimensional object, our 7-by-7-by-7 unit cube, is bounded by a surface area of 294 square units.

Conserving the boundary

We can start with the same rule that we started with when exploring two-dimensional space. We conserved the perimeter of our square – the boundary of the square. So now, we will conserve the boundary of our cube, its surface area. And we are going to perform the same operation as before. We are going to squash the object to see what happens. Before, we squashed the square from above reducing its height. So now, we will squash the cube from above, reducing the height of the cube.

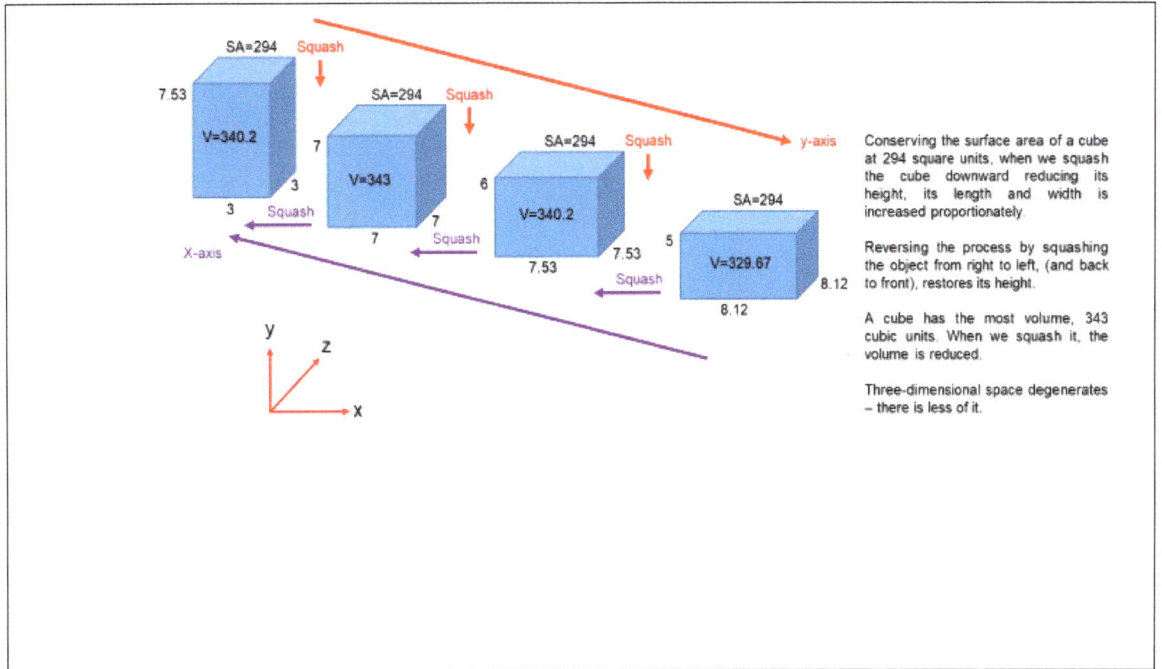

Figure 4.

That a three-dimensional space is defined uniquely by its volume, allows us to describe a reduction in volume as a degeneration of the dimension and an increase in volume as a generation of the dimension. After degeneration, there is less volume; there is less three-dimensional space. After generation, there is more volume; there is more three-dimensional space. We can see that maintaining a constant surface area for its boundary has no conserving effect on the amount of volume there is when we squash the cube or when we reverse squash it.

In this exercise, if we keep squashing down on the cube, the height will keep reducing. Eventually the height will be zero. At that point, we no longer have a three-dimensional object. We have a rectangle. A two-dimensional plane. And yes, the plane will have an area of 294 square units. Because that was our rule: to maintain the surface area boundary at 294 square units. But having squashed the height to zero, we no longer have a three-dimensional object. So, what is the 294 square units the boundary of? Clearly, we have destroyed our three-dimensional object and so surface area is no longer a boundary. It no longer *bounds* three-dimensional space. Having destroyed the cube, we can't simply reverse squash the remaining rectangle to recreate it. If we reverse squash a two-dimensional object, we may alter its area. But we can't reverse squash a two-dimensional area and generate a three-dimensional volume. Once the cube is gone – it's gone for good.

Conserving the dimension

We observe that when we squash our three-dimensional object but conserve its surface area boundary, its width and length measures increase. But, again, if we keep squashing, eventually the height will be zero and we end up with a two-dimensional object. The three-dimensional object is destroyed and can no longer be restored to a cube by pushing it back into shape. So again, maybe we made up a wrong rule? Could we make up a different rule, so that we can squash our object but never destroy it? Turns out, we can use the same rule we used with two-dimensional space. We can conserve the volume of the object; conserve its three-dimensional space when we squash it and see what happens.

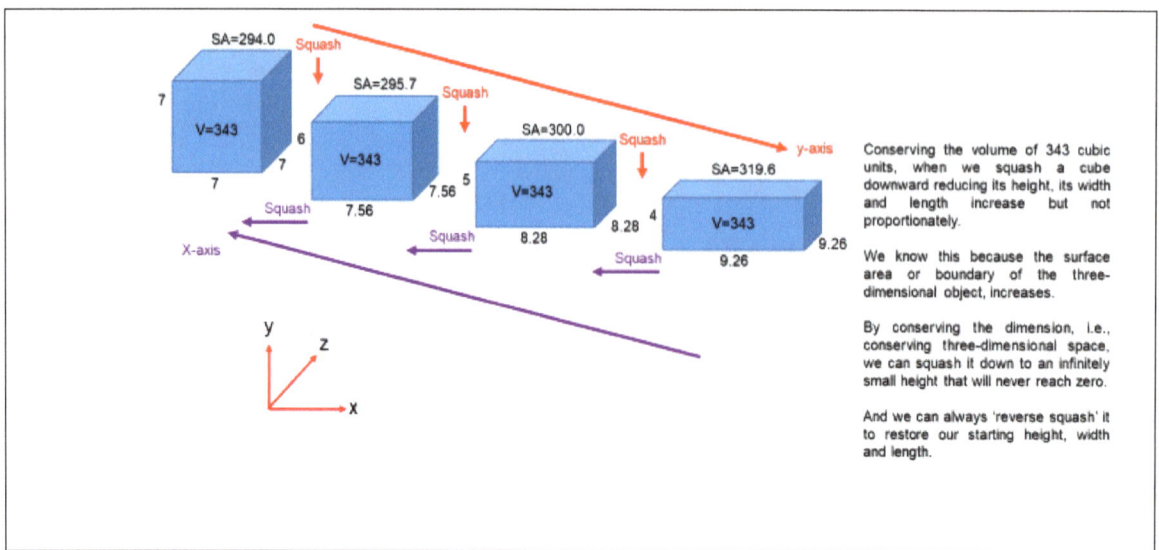

Figure 5.

Our rule works – again! Squashing the object does not destroy it. We get longer and longer length and width values, but the height of the object can never reach zero. Our rule forbids it.

And it's still clever. We can squash the height of what was our cube, down to an object having near infinite width and length. That's because by conserving our three-dimensional space, our 'volume', we can squash the height down, for ever getting closer to zero but never reaching it. If we reached it, our volume would become zero. So, we are only allowed to make the height smaller and smaller and smaller still. And when we do that, conserving our three-dimensional space of 343 cubic units, the length and width get longer and longer and longer.

And, again, we can 'reverse squash' our object and recover our original height, width, and length. That's because we always have a three-dimensional object to 'reverse squash'. It is never squashed down to a two-dimensional plane. With our new rule, if we have a three-dimensional space, squashing it can't destroy it. It doesn't degenerate.

Deformation of three-dimensional space

If we change the shape of a three-dimensional object, allowing the size of its boundary to alter to maintain the amount of space bounded, we can look at the effect on its boundary, its two-dimensional surface area.

Take a cube with edges each 10 units in length. The volume of the cube is the product of width, height, and length, 1000 cubic units. The perimeter of the cube is the sum of the surface areas of its six faces, 600 square units. Then we squash the cube so that its height is reduced from 10 to 5 units, but we want to maintain a constant volume value of 1000 cubic units. The reduction in height gives way to increases in width, length, or both.

We can calculate that to squash the cube down to five units high, the width and length must each be the square root of 200 units (because volume is 1000, and 200 x 5 = 1000), or approximately 14.142 units. A simple check: 14.142 x 14.142 x 5 = 999.98, approximately our original 1000 cubic units. Use the square root of 200 in place of 14.142 to check the result exactly. If we now check the value of its boundary perimeter, we can calculate that it is approximately 683 square units. That's the sum of the surface area of the two 14.142 x 14.142 top and bottom faces, plus the sum of the four 5 x 14.142 side faces of the cube (which strictly speaking is now called a cuboid).

By squashing the cube, we have increased its boundary from a surface area of 600 square units to 683 square units. Volume and surface area are attributes of a three-dimensional space. Squashing the cube can be thought of as changing the shape of the three-dimensional space but not its spatial value, its volume. That a three-dimensional space is defined uniquely by its volume allows us to associate a reduction in volume with degeneration and an increase in volume with generation. However, neither occurred. We kept the volume constant at 1000 cubic units. We used our rule!

But the cube's surface area did increase – from 600 to 683 square units.

So, we find that deforming a three-dimensional volume increases its two-dimensional surface area, its boundary.

Applying the same reasoning, we would arrive at the proposition that an increase in surface area would represent a generation of the dimension, a generation of two-dimensional space. By squashing a cube and conserving its volume, we have increased its two-dimensional boundary. By deforming a spatial dimension, we have increased the magnitude of its next lowest dimension – its boundary.

Now, we should take a short break here to consider the obvious. And as frequently happens, the obvious is sometimes overlooked. The way a spatial dimension maintains its defining attribute when dealing with a deformation at its boundary is to stretch the boundary. The boundary stretches and the space is conserved. And that … implies pressure … which in turn implies force … and hence energy is involved, (remember energy?). A deformed two-dimensional object maintains its area, and we are compelled to believe this happens because the object is pressurised. So when a force pushes against its bounding perimeter deforming it, the internal pressure resists compression of the space, causing the boundary to stretch instead. Noteworthy is that it would seem then that the higher the spatial dimension, the higher the pressure (conserving its defining attribute – its space). Areas stretch perimeters, volumes stretch surface areas. It never goes the other way around – in the other direction. So we might tinker with the idea that a spatial dimension, by being pressurised, exerts a force at its boundary, able to ward off attempted intrusions. That outward force deals with attempted intrusions by stretching the boundary. The object's shape may change, but its space is always conserved. And this is something that we will come back to later because it offers a surprisingly *useful* perspective.

We can say then that deforming a dimension means changing its shape to a less economical one. And doing that deforms its boundary – the next lower dimension – because it will change *its* shape to a less economical one. When we started with our economically shaped three-dimensional cube, its six faces were all squares – economical shapes. When we squashed the cube, the four sides of the cube became rectangles – less economical shapes. The shape of the top and bottom faces remained squares. So taken as a whole, the surface area is then made up of two shapes that are economical and four shapes that are not. The shape of our surface area was changed to a less economical shape when we deformed the cube.

Let's redo this idea with a sphere. To deform a sphere, we need to either pull or push it out of shape but conserve its volume. A sphere is already three-dimensional space in its most economical shape. Its surface area is two-dimensional space. When we change the shape of the sphere but conserve its volume, we can't avoid increasing its two-dimensional boundary. We end up with a shape that is not quite a sphere having a surface area somewhat greater than the surface area we had before we changed the sphere's shape.

Four-dimensional Space

From a very early age, I recall learning that time was the fourth dimension. I also recall never being able to grasp that concept. I had a good understanding of spatial dimensions, which developed as I grew up and it always irritated me that after spatial dimension three, we had to accept time as the fourth dimension. It was not until years later, after spending some time reading relativity theory, that

I learned how it was that time acquired its dimensional status. Unfortunately, that didn't convert me. I had the idea in my head that there should be a fourth spatial dimension and set out to discover it. I also knew that if it were discoverable, and discovered, then I would need to revisit the concept of time. The science ascribing the fourth dimension to time was well established in the context of our universe being constructed as a spacetime continuum. Any discovery of a fourth spatial dimension would require an assessment of its place in the cosmos relative to that.

I learnt at university a little about the requirements for moving from one spatial dimension to the next. The rules are basic. The axis of every additional spatial dimension needs to be at right angles (90 degrees) to the axis of any other dimension. If we say that the first dimension has a left-right axis (x), then the axis of the second dimension must be at right angles to that. So, if we say the second dimension has an up-down axis (y), then the axis of the third dimension must be at right angles to both the x-axis and the y-axis. And if we say the third dimension has a forward-backward axis (z), then the axis of the fourth dimension must be at right angles to the x-axis, the y-axis, and the z-axis. Can you see the problem?

The problem is that because of the way we experience our universe, after up and down, left and right, forward and backward, there are no directions remaining that present at 90 degrees to every other direction. The only direction I could dream up was inward and outward (w), not unlike the sense one has when one thinks about being inside one's mind; that would be inward. Now in a physical context, outward would be analogous to poking a finger into the air and see it disappear into a hole that leads somewhere that is not in our universe; and inward would then be the direction taken as the finger is pulled out of the hole, back into our universe, until it was fully visible again. A childish concept to be sure but it served the purpose and provided what I needed to carry on.

We could always make drawings of objects having more than three dimensions on paper. Today, you can go on the internet and find drawings of cubes with four, five, and more dimensions. I was attempting to understand how a fourth dimension might fit *into* our universe – not a drawing of what a four-dimensional object might look like. But I had to start with a drawing in order to understand the rules.

Innovating the hypercube

Having established at least the concept of a fourth spatial dimension, I set about exploring how four-dimensional objects would look when represented on a sheet of paper. After all, there were rules governing the representation of one-, two- and three-dimensional objects on paper, so all that was needed was to identify those rules and extend them to include the representation of a four-dimensional object. The rules are simple. There are three directions in our universe, (x, y, z) that is, left and right, up and down, backwards and forwards. Draw an object of any dimension. Draw a second object of that dimension and move it a distance from the first one in a direction that is at right angles to all directions used when drawing the object. Then join every point of the first object to every point of the second.

Take a point on a page. It is a zero-dimensional object. Take a second point on the same page at a distance from the first. We can use any direction since no direction was used to draw the point. Join the two points, that is, draw a line between the two points. We have drawn a line which is a one-dimensional object. One dimension higher than the zero-dimensional points that make up the line. Take note that one of the three directions (x, y, z) was used up when we drew it; let's say it was 'x'.

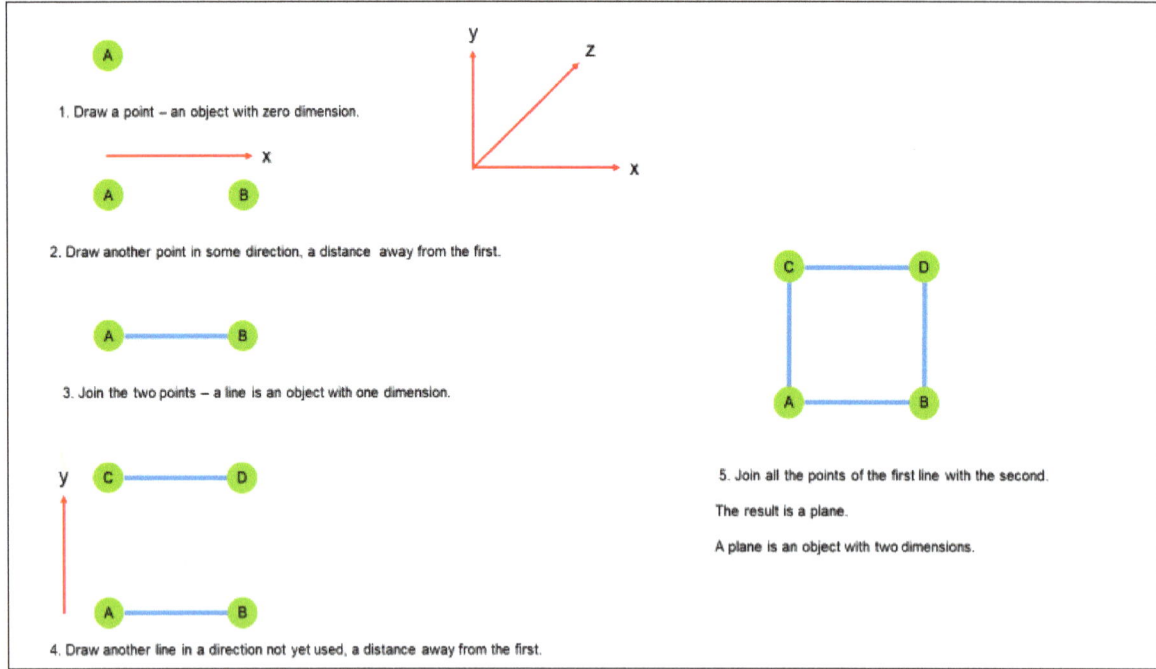

Figure 6.

Draw a second line on the same page at a distance from the first and in a direction that is not 'x'; let's use 'y' this time. Now join every point on the first line to every point on the second line. We have drawn a plane which is a two-dimensional object, one dimension higher than the one-dimensional lines that make it up. We have used up the x and y directions.

Draw a second plane on the same page at a distance from the first and in a direction that is not 'x' or 'y'; we are compelled to use 'z', the only remaining direction from our set of directions {x, y, z} so let's use 'z'. Now join every point on the first plane to every point on the second plane. We have drawn a cube, which is a three-dimensional object, one dimension higher than the two-dimensional planes that make it up. We have used up the x, y, and z directions.

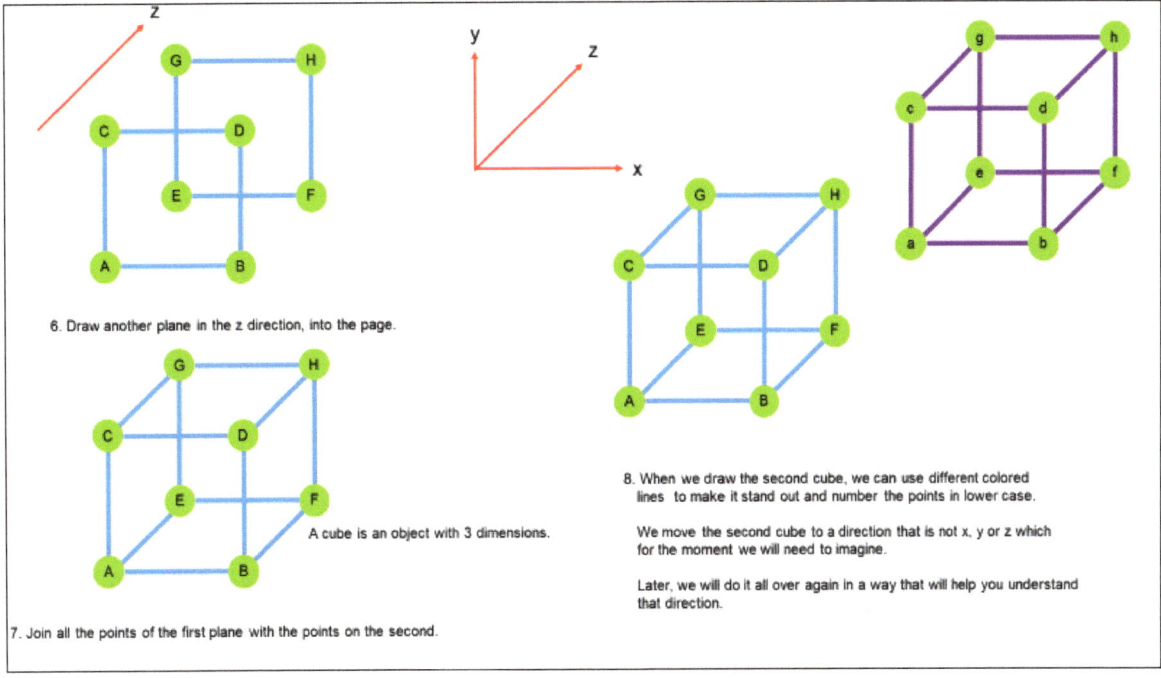

Figure 7.

Draw a second cube on the same page at a distance from the first and in a direction that is not 'x', 'y' or 'z'. Let's call that direction 'w'. Now join every point on, (and inside), the first cube to every corresponding point on, (and inside), the second cube. We have drawn a hypercube which is a four-dimensional object, one dimension higher than the three-dimensional cubes that make it up.

Clearly, there are an infinite number of points on any line or plane, or on and in any cube, so following these rules precisely won't result in anything recognizable on paper. And of course, we would not be physically able to join every point from one object to every corresponding point on the other since there are an infinite number of points. All we want to do is to sketch an outline of the object and to do that we only need to join the end points.

By labelling the first point A and the second B, gives us the line (A, B). By labelling the second line (C, D) enables us to create the plane by joining A to C and B to D, so that we have the outline of a square labelled clockwise (A, C, D, B). In like manner, if we label the second plane (E, G, H, F) then we can join A to E, B to F, C to G, and D to H – describing the outline of our three-dimensional object, the cube. When we draw the second cube we can label it in lower case, (a, b, c, d, e, f, g, h). Now join every point on (and inside) the first cube to every corresponding point on (and inside) the second cube.

Total Relativity and Dimensional Dynamics

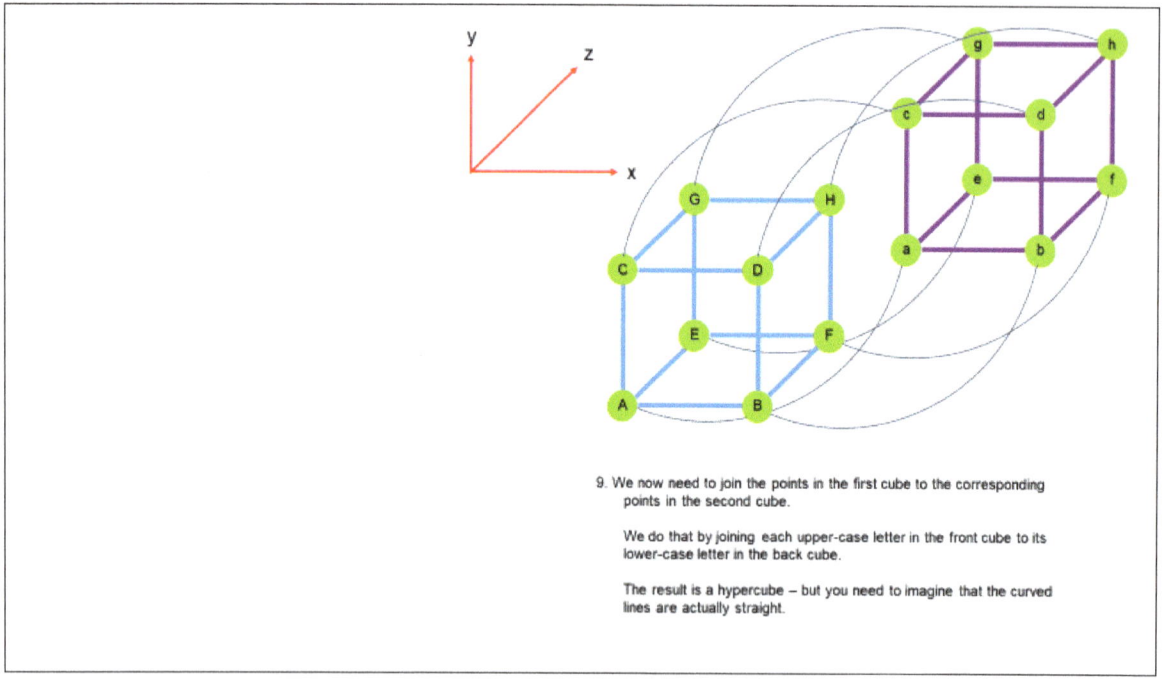

Figure 8.

Of course, we are only interested in the outline and so we join upper case letters to lower case letters, that is, A to a, B to b, and so on. If you did what I originally did, you would have simply drawn a second cube on the page and joined upper case to lower case letters. You would have done this because just as we simulated the direction 'z' going into the page on the two-dimensional paper we're using, we would have, by extension, simulated the fourth direction 'w' as well.

Once the corresponding corner points of the two cubes are joined, it would be a little difficult to imagine the resulting diagram is the same as the now much-publicised diagrams of a hypercube.

I had exactly that difficulty; I couldn't easily see that a hypercube constructed in this way was the same as the diagram of the hypercube found in popular science books. My hypercube didn't even have straight lines connecting the two cubes. But I wanted you to see what I ended up with by following the rules. And to understand that what was important to me was to imagine the fourth direction to be real – to exist – but in a way that we could not directly experience it. We cannot look in that fourth direction, nor travel in it. Like poking a hole in the air – that's where it is – but only if we imagine it so.

To see that it is indeed the same diagram as we see these days in popular science books, reconstruct the rear-most cube so it is smaller than the foremost cube. Then simply pull the smaller cube into the larger one and make the connecting lines straight. What is interesting about this way of presenting a hypercube is that it brings to life the concept of the fourth direction 'w' as being an inward-outward one. You have a smaller cube in a direction that is inward from the larger one. Or you can say that you have the larger cube in a direction that is outward from the smaller one. See how that works?

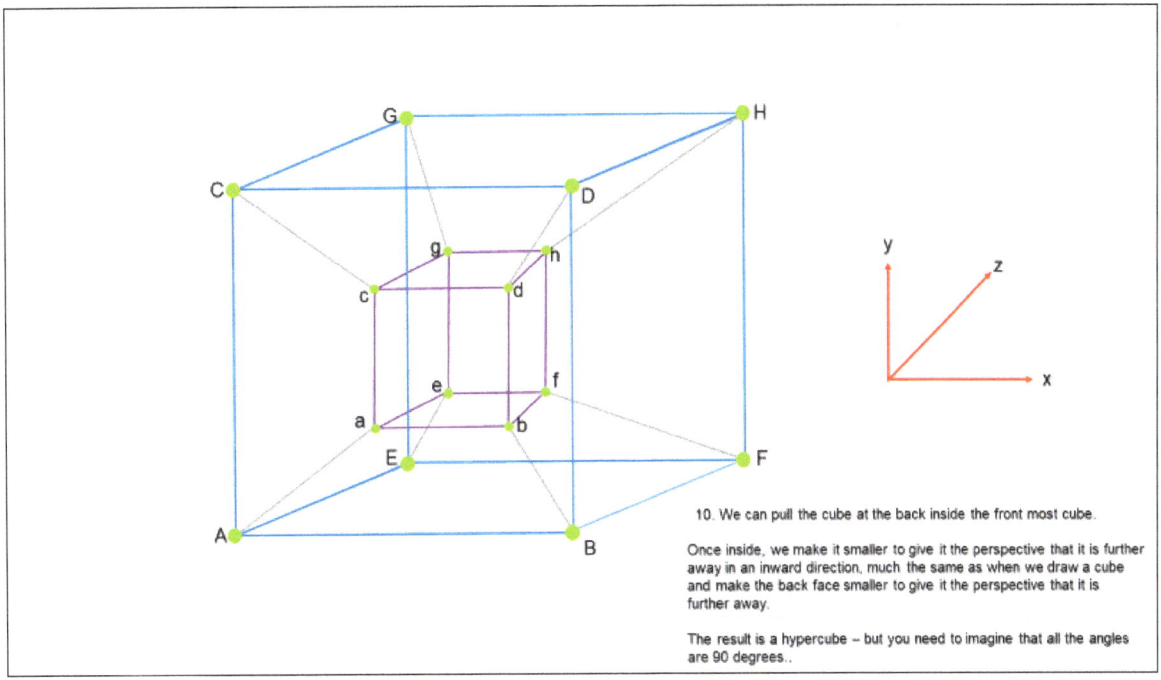

Figure 9.

This simplistic approach reveals that an extension of rules is all that is needed to depict an object of a particular dimension as an object of a higher dimension.

The same rules we used to construct a line from two points, a plane from two lines, and a cube from two planes – were used to construct a hypercube from two cubes; and the only difficulty experienced was how to represent the resulting image on two-dimensional paper.

When we draw a three-dimensional cube on paper, we sometimes make the back face of the cube smaller than the front face to provide a sense of depth to the drawing. In our hypercube drawing, reducing the size of the rear most cube, then repositioning it within the larger foremost cube is a similar process. As with a three-dimensional cube we imagine its rear face is further away, with a hypercube we imagine its smaller cube further away but inside the larger one. Just as with the three-dimensional cube drawing, we imagine the rear face being in the 'z' direction from the front face - with the hypercube we imagine the smaller cube is in the 'w' direction, an inward direction from the larger cube. We can imagine now that the direction chosen as being other than from the set {x, y, z}, is indeed inward; or from the perspective of the smaller cube viewing the larger one, an outward direction.

Deformation of four-dimensional space (varying the three-dimensional boundary)

Before we look at the deformation of a four-dimensional object, we would benefit from taking some time to understand our hypercube, identify its perimeter, and make some defining claims about four-dimensional space (which we will call *hyperspace*). For context, we call one-dimensional space a line, two-dimensional space an area, three-dimensional space a volume and now, four-dimensional space a hyperspace.

We can colour the edges of our hypercube in a way that allows us to recognize where an extracted element of the hypercube's bounding volume came from after we extract it.

First, imagine we have a drawing of a cube, and we want to show separately its top, bottom, or one of its four sides. These would each be *extracted elements* of the cube's surface area – its boundary. But they would each look the same. They are all squares. After we extract one of these squares, how to tell if *that* square came from the top, bottom, or one of its sides? Well, of course, if the points of the cube where the edges meet are labelled with letters, that would tell us.

But with a hypercube, there are more labelled points, and we are working with a fourth dimension: the w-dimension. So, it can get quite difficult to look at a piece of a hypercube and see where it came from. Colour helps us to do that. We will also need to pay special attention to these cubes – the cubes that comprise the boundary of the hypercube – its volume perimeter – to understand how when we squash the hypercube, where we are squashing, and in which direction. It was relatively simple when dealing with less than four dimensions. But with four dimensions, more is demanded of us to comprehend what happens when we squash the hypercube – when we deform a four-dimensional object.

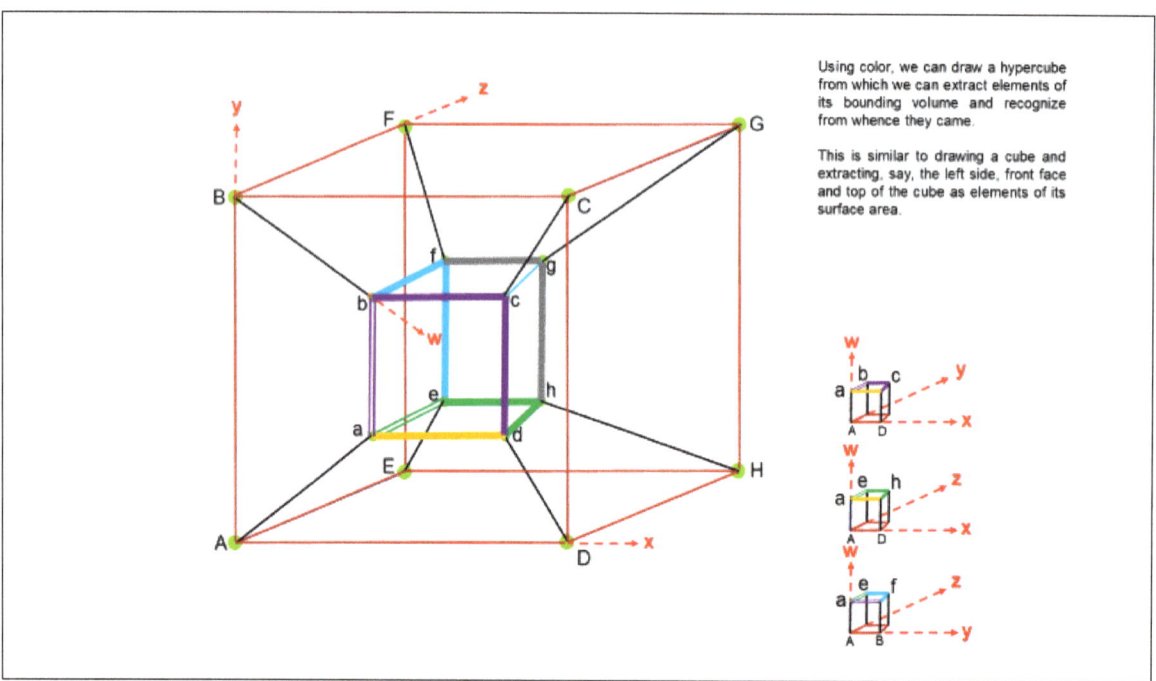

Figure 10.

From the side charts of our hypercube diagram, where we abstract three of the four axes at a time, we can identify those points where the value of the *w*-coordinate is zero, that is, where the w-axis intercepts the y-axis and the z-axis. These points exist in our universe; on the xy, xz, and yz real number planes, in our xyz spatial universe. The points 'A' through 'H' describe the cube that exists in the xyz universe. The points 'a' through 'h' describe the cube that also exists in our universe but on the other side of the hyperspace that these cubes bound – much the same as the back face of a cube is on the other side of the volume the cube's faces bound. Our rules tell us that the perimeter of hyperspace is a volume, just as the perimeter of volume is an area and the perimeter of area is a line; always one lower dimension.

Just as we can calculate the perimeter of a cube's volume by summing up the area of its six faces, we can calculate the perimeter of a hypercube's hyperspace by summing up the volumes of its bounding cubes – the volume perimeter. The perimeter of our hypercube is the sum of the volumes of the inner and outer cubes which we know from our side chart are one cubic unit each. Recall that we earlier established the angles ABb, BAa, etc., as right angles in the hypercube and we use the side charts to show that; they just don't look like right angles, but neither do DAE, ABF, etc., which are depictions of a three-dimensional object on two-dimensional paper – but we are perhaps more accustomed to identifying these as right angles.

We can now also identify the size of the hypercube boundary as being the sum of the 'volumes' of the six projected cubes – one cubic unit each, plus the outer cube and the inner cube, summing up to eight cubic units. That's eight units of volume bounding one unit of hyperspace.

The analogy is that of the boundary of a cube. A cube can be generated from a single area – its front face. When we take a copy of the front face and move it backward to form a cube, the cube's sides, top and bottom are generated. The sum of the sides, top and bottom, as well as the original front face and the final back face, form a continuous surface area – the perimeter or boundary of the cube.

But there is something significant here to note.

Just as the surface area of a volume is not used to calculate how much volume we have, the volume perimeter of a hyperspace cannot be used to calculate how much hyperspace we have. So, just as the surface area of a volume does not define the quantity of volume an object has, nor does the volume of the *volume perimeter* of a hyperspace define the quantity of hyperspace a hyper-object has. But this is a special case. Having created the hyperspace using a one-unit cube projected one unit in the w-direction, we can say that produces one unit of hyperspace.

What we are describing here is that a hyper-object, a hypercube in this instance, *contains* hyperspace (four-dimensional space), that is *bounded* by a volume (three-dimensional space). Note that the configuration of the boundary defines the object, and it is the object that *contains* the space. If we chose not to depict an object at all, we can say that hyperspace is bounded by volume, and more specifically, that the volume *does not* contain hyperspace but bounds it. To make this a little more comprehensible, we can drop down one dimension and say: if we choose not to depict an object at all, we can say that a volume is bounded by a surface area, and more specifically, that the surface area does not contain volume but bounds it.

Having defined the volume perimeter of our hyperspace, we can again refer to our rules and note that when we squash our hypercube while keeping its defining attribute constant – its quantity of hyperspace constant – our expectation is that the volume perimeter bounding the hypercube will increase. Because we are dealing with four dimensions depicted on two-dimensional paper, we need to pay close attention to fully appreciate how it is that we can determine what to squash and in which direction, such that the rules are obeyed, and the deformation of the hyperspace is achieved, resulting in a *generation* of the volume perimeter – an increase. As we shall discover, this task is not intuitively obvious.

When we squashed the square into a rectangle, we pushed in the negative direction of the y-axis. Alternatively, we could have pushed in the negative direction of the x-axis. And we could have pushed in the positive direction instead. While keeping the area constant, we allowed the perimeter to change size to accommodate and conserve the area – we stretched the perimeter of the square, its one-dimensional boundary.

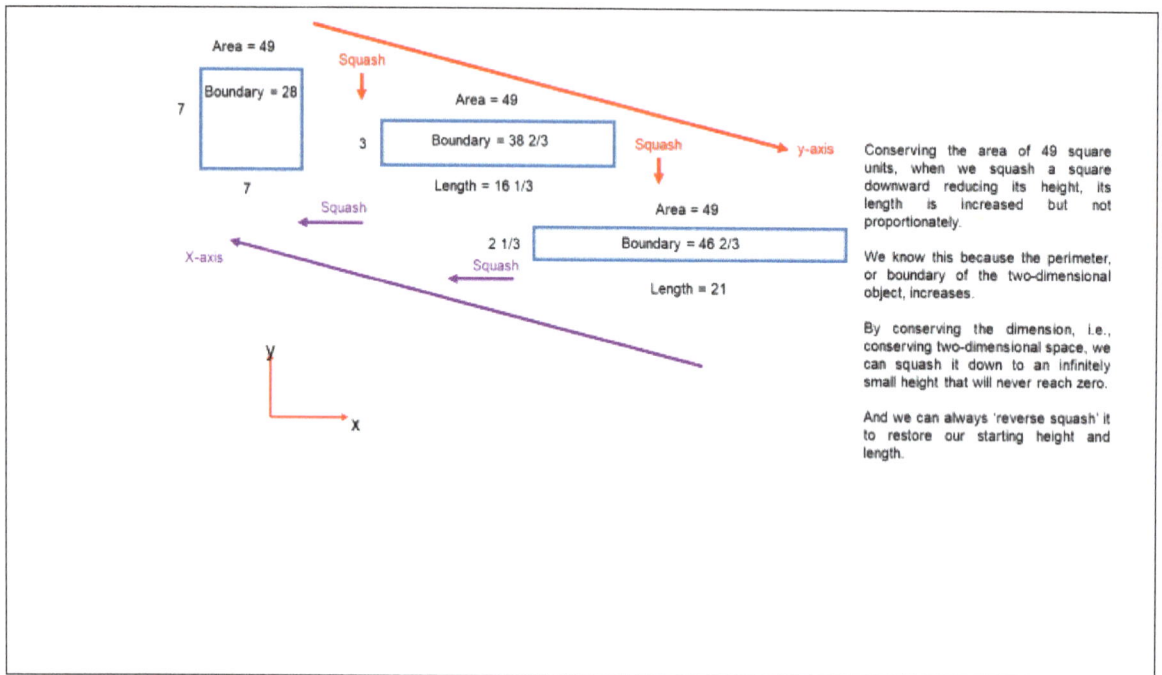

Figure 11.

When we squashed the cube, we pushed in the negative direction of the y-axis. Alternatively, we could have used the x-axis, or the z-axis. And we could have pushed in the positive direction. While keeping the volume constant, we allowed the surface area to change size to accommodate and conserve the volume – we stretched the perimeter of the cube, its two-dimensional surface area.

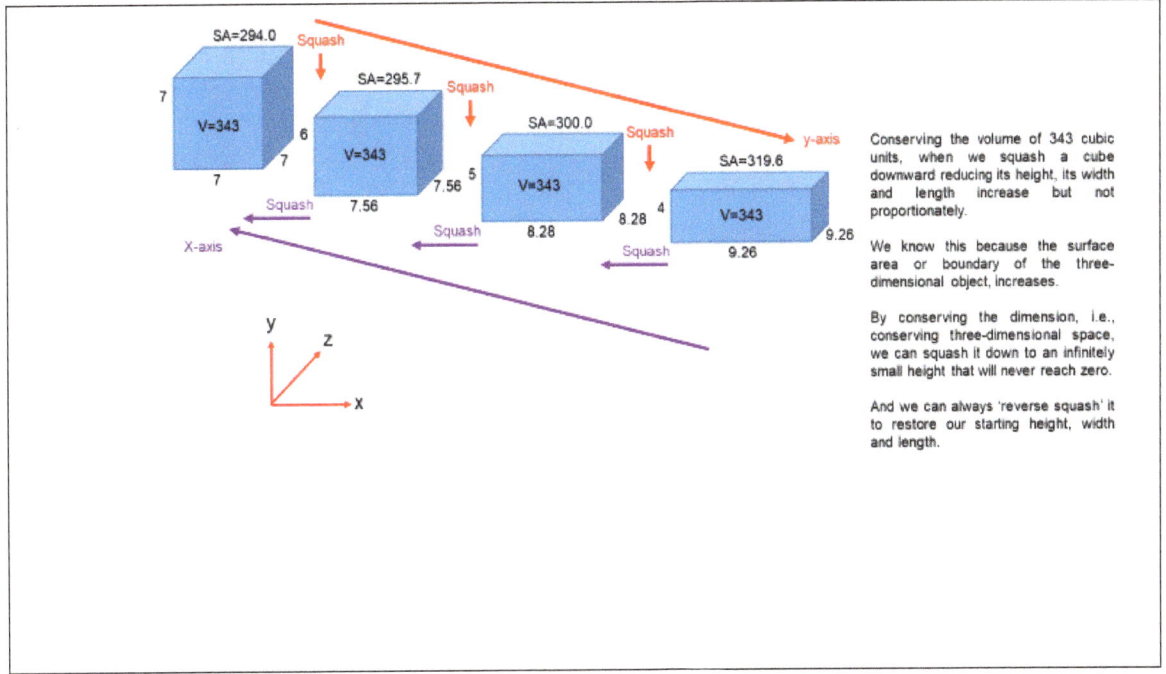

Figure 12.

We now want to squash a hypercube. There's a simpler way to do this than thinking about which axis and in which direction to squash. We used axes and directions with two- and three-dimensional objects to help us understand what we were doing and why. It allowed us to make some calculations. We now know what we are doing and why, so we can remove some of the complexity used earlier.

If we think about how we *generated* a square from a line, what we did, is we took a copy of a line and 'moved' it away at right angles to the directions used to draw the line. Let's call the original line line-1 and call the copy we moved line-2. We generated the two-dimensional square from a one-dimensional line by moving line-2 away from line-1. Then we joined all the points on line-1 with the corresponding points on line-2. OK, we have our square now, our two-dimensional space. We can squash our square by simply moving line-2 backward toward line-1. But as we do this, we must apply our rule and conserve the surface area – its two-dimensional space. Of course, when we do this, the perimeter of the square will stretch.

Similarly, we *generated* a cube by taking one face of a cube – a square – and moving a copy of it away. If we call the front face of the cube face-1 and we move its copy, face-2, into the page, that will generate the cube for us when we join all the points on face-1 with the corresponding points on face-2. We can squash the cube by simply moving face-2 back toward face-1. But as we do this, we must apply our rule and conserve the volume – the cube's three-dimensional space. Again, when we do this, the surface area of the cube will stretch.

I think you're getting the idea. To generate our hypercube, we moved the copy cube-2 in the w-direction away from the original cube-1. So, all we need to do to squash the hypercube is to move cube-2 back toward cube-1 along the w-axis whilst conserving its hyperspace – its four-dimensional space. When we do this, we expect the volume perimeter bounding the hyperspace to stretch.

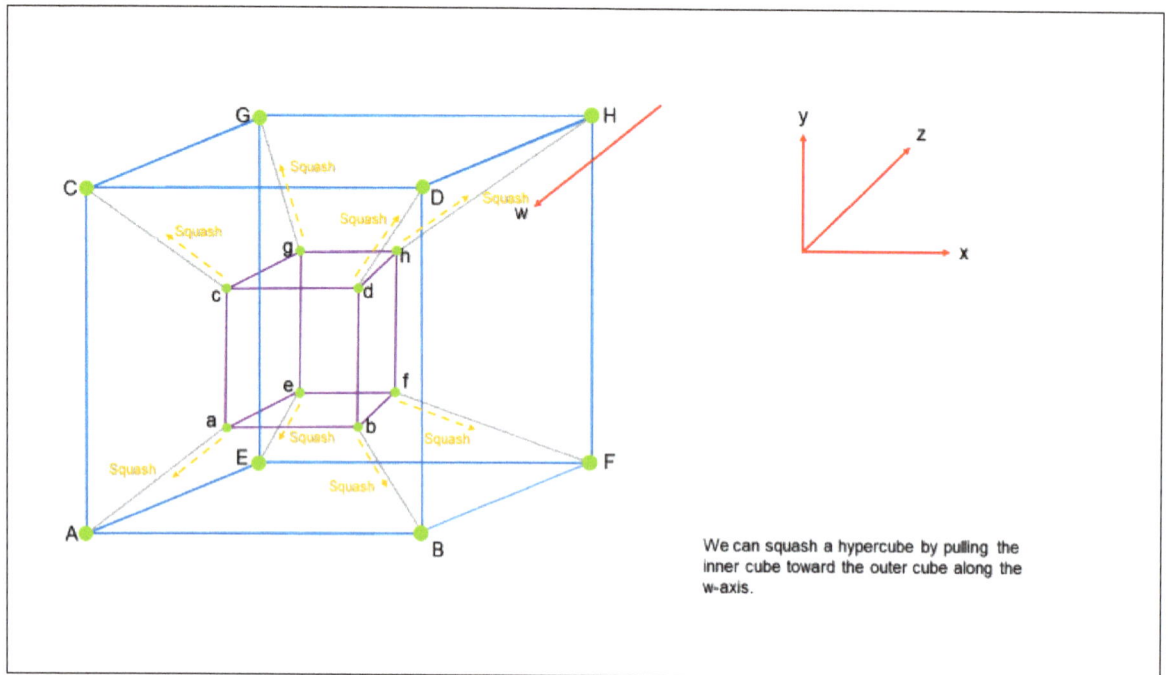

Figure 13.

Of course, when we squash the hypercube, we can push in any direction of the x-axis, the y-axis, the z-axis, or the w-axis. To do it, it's just easier to pull cube-2 back toward cube-1, because when we do that, we can imagine the inner cube, cube-2, getting larger as it comes toward us from the w-direction to cube-1. As the separating distance in the w-direction between the inner and outer cubes reduces, the hypercube's perimeter, its bounding volume must increase – our rule tells us this must happen. The volume of some or of all the inner, outer, and projected cubes must increase to keep the quantity of hyperspace constant; to maintain the prime property of the hyper-object. To conserve the hyperspace.

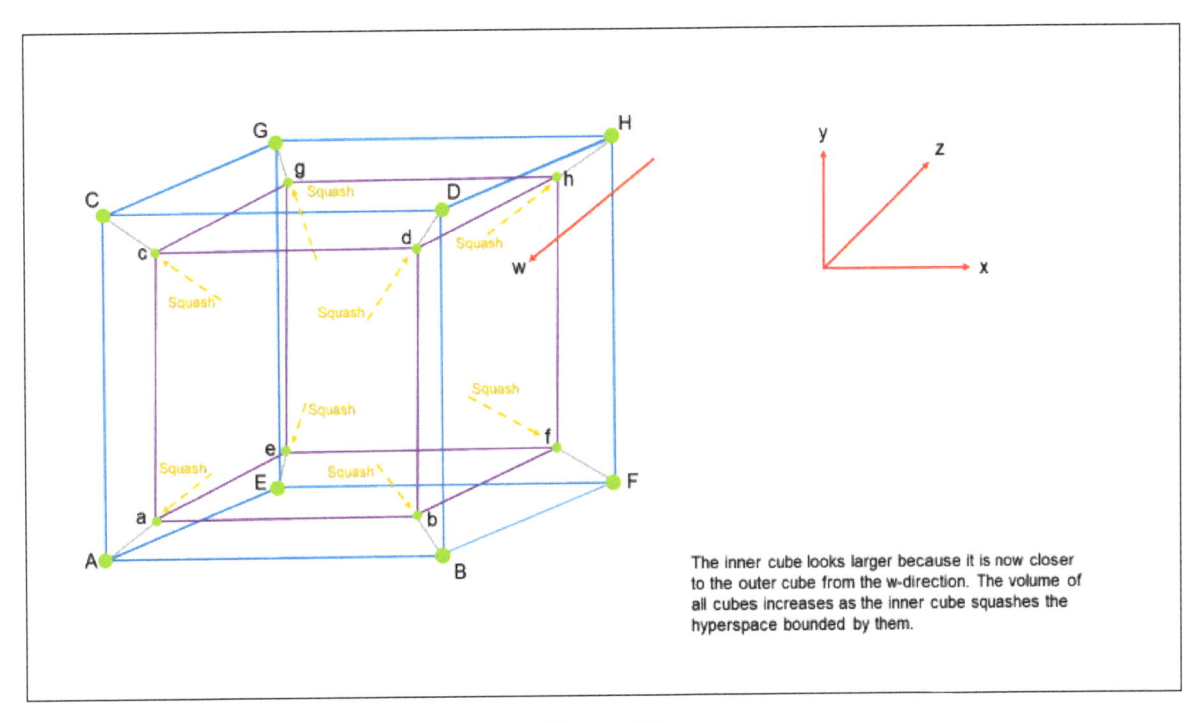

Figure 14.

This simply demonstrates that a distortion of a hyperspace results in an increase in its volume perimeter (an increase in the volume of our three-dimensional universe?). Continue to squash the hypercube and as the distance between the inner and outer cubes approaches zero, the sum of the volumes of the cubes bounding the hyperspace approaches infinity.

Interim conclusion

We started with the idea that 'space' might be *something* rather than *nothing*. We explored how four spatial dimensions might define an object, such as a hypercube. We thought about how a hypercube might contain space and we named *that* space - 'hyperspace'. We demonstrated how hyperspace could interface to the three spatial dimensions we are all familiar with. The three spatial dimensions of our universe would form the boundary of the hyperspace. We showed how distorting a hyper-object inflates its boundary, its volume perimeter, and possibly inflate it to a size approaching infinity.

We can extend these ideas when we think about the expansion of *our* universe. We can think about our universe in terms of an expanding volume perimeter of a hyperspace. We can infer that the expansion we see in our universe need not result exclusively from a Big Bang *event – from a singularity*. That perhaps the fabric of space itself is stretching in response to a deformation, a 'squashing', of a hyperspace, so that its bounding volume perimeter – our universe? – expands. We measure the expansion by measuring things *in* space. Galaxies move further and further apart. But what if it really is space itself that is expanding? Imagine that as possible … and you can imagine TRADD.

Expanding space

The measurements taken by science tells us that our universe is expanding. Popular science publications explain how now, different measurements, provide evidence for the expansion of space. It is our task here, to take what we have learned about 'dimensional dynamics' and apply that to the notion of an expanding universe – of expanding space.

We have seen how, when conserving a dimension whilst deforming it, its next lower dimension expands. Deform an area and its perimeter expands. Deform a volume and its surface area expands. Deform a hyperspace and its bounding volume expands.

Our universe has three spatial dimensions – the ones we are familiar with. The standard model adds a fourth dimension, calls it 'time', and then describes our universe as an unbroken expanse of continuous space and time – calling it the spacetime continuum. But what if we dispensed with 'time' – at least temporarily? What if we instead looked at how the remaining three dimensions of our universe – its *'volume'* – might expand if it were the boundary of some four-dimensional space? And as we have shown, deforming a four-dimensional space can result in an expansion of its three-dimensional bounding *volume*. If our universe is the bounding *volume* of some four-dimensional space – *a hyperspace* – that is being deformed, that would result in our universe expanding.

How might space expand?

It should be easy for us now to imagine how a three-dimensional volume might expand, resulting from the deformation of a four-dimensional space. But when we went through how that might happen, we used an object called a hypercube because it was easier to visualise and understand. We also learned that the most economical way to enclose a volume was in a sphere. So, to begin with, let's decide, for the sake of our exercise, that our universe is a sphere – or perhaps the bounding volume of a hypersphere. When I say, 'our universe', I'm talking about the three dimensions of space that we all know and love and are familiar and comfortable with.

For the moment – at least – we have dispensed with 'time'. But we have not dispensed with the w-direction. Remember the w-direction? It's the direction that we can move a volume in, to generate a four-dimensional object. It's how we generated our hypercube. We moved a copy of a cube in an inward direction that we called the w-direction. After we created our hypercube, we decided to squash it – but conserve the hyperspace it contained – and as a result, the volume bounding the hyperspace expanded.

If we say our universe is the bounding volume of a hypersphere, then we can use the idea of a sphere to generate a hypersphere. All we need to do is create a copy of a sphere and move it in an inward w-direction in the same way as we moved our copy of a cube to generate a hypercube. And after we do that, we can imagine that we have a picture of a small sphere inside a larger sphere. The smaller sphere is not actually smaller. It's just further away from the larger sphere in the w-direction. We just drew it smaller to give it the perspective of being distant.

Having created our hyperspace inside our hypersphere we then want to squash the hypersphere but conserve its hyperspace. When we do that, we know what will happen. The volume bounding the hyperspace will expand. And the volume bounding the hyperspace is the volume that we recognise as the large sphere, the smaller sphere that we pushed in the w-direction, and the projected spheres which we will have trouble imagining a drawing of, but that we know must exist. I guess now you'll appreciate why we started with a cube – because the projected cubes in the hypercube can more easily be drawn, comprehended, and measured.

You will recall that when we squashed the hypercube, we pushed the inner cube in the negative w-direction. We pulled it back toward the outer cube. We did that so it would be easier to explain what happened when the 'smaller' *distant* cube was pulled toward the *closer* 'larger' cube, which was the large outer cube in our drawing. But did you figure out that we could just as easily have squashed the hypercube in the positive w-direction? To do that, we would have needed to push the outer cube in the w-direction, closer and closer to the inner cube. The result would have been the same; the bounding volume of the hypercube would have expanded.

We didn't do it that way because it would have been harder to show what was happening. We were limited by a drawing on two-dimensional paper. We would have had to move the outer cube further away in the w-direction, closer to the inner cube. And to show that, we would have had to draw the outer cube getting smaller and smaller to give it the perspective of ever-increasing distance. Distance away in the w-direction.

But if we then squashed the hypercube, this time in the positive w-direction, the volume of the cube that we recognise as the outer cube would have expanded. (The volume of the inner cube would have expanded, too). But to show something expanding on paper, we would normally make it bigger. See the problem? To squash the hypercube in the positive w-direction, we would have needed to draw the outer cube smaller and smaller to show that we were pushing it away, but we also would have needed to draw it bigger and bigger to show it was expanding. That would have been a little confusing for some of us to see what was really going on; it would have been difficult to describe in a drawing.

But now we have a hypersphere. And we want to squash it. And we *do* want to squash it by pushing the outer sphere in the positive w-direction. So rather than pull the inner sphere back in the negative w-direction, we instead want to push the outer sphere in the positive w-direction – toward the inner sphere. And when we do *that*, whilst conserving the hyperspace in our hypersphere, we should expect the volume bounding the hyperspace to stretch – to get bigger – to expand.

Now just as we know the large cube in our hypercube expanded when we pulled the smaller cube in the outward direction (in the negative w-direction), we should see the large sphere expand when we *push* it in the positive w-direction toward the inner sphere. All spheres of the hypersphere can expand as we squash the hyperspace bounded by them. The inner sphere, the outer sphere, and the projected spheres.

Suppose our universe is the volume bounding a hyperspace. This volume would consist of the sum of the volumes of the inner, outer, and projected spheres. Just as the six faces of a cube form one continuous surface area — analogous to the continuous surface area of a sphere — the inner, outer, and projected spheres of the hypersphere create one continuous volume. This volume, which is now expanding, represents the space in our universe.

We have shown that if our universe were the bounding volume of some four-dimensional space – a hyperspace – then if that hyperspace was being deformed (squashed), our universe would expand.

So, what is the TRADD model telling us? What have we got so far?

Our universe is a volume bounding a hyperspace. Some of that volume, perhaps in the shape of a sphere, is moving in the w-direction. It is squashing the hyperspace. Because of that, the entire volume bounding the hyperspace – our universe – is expanding.

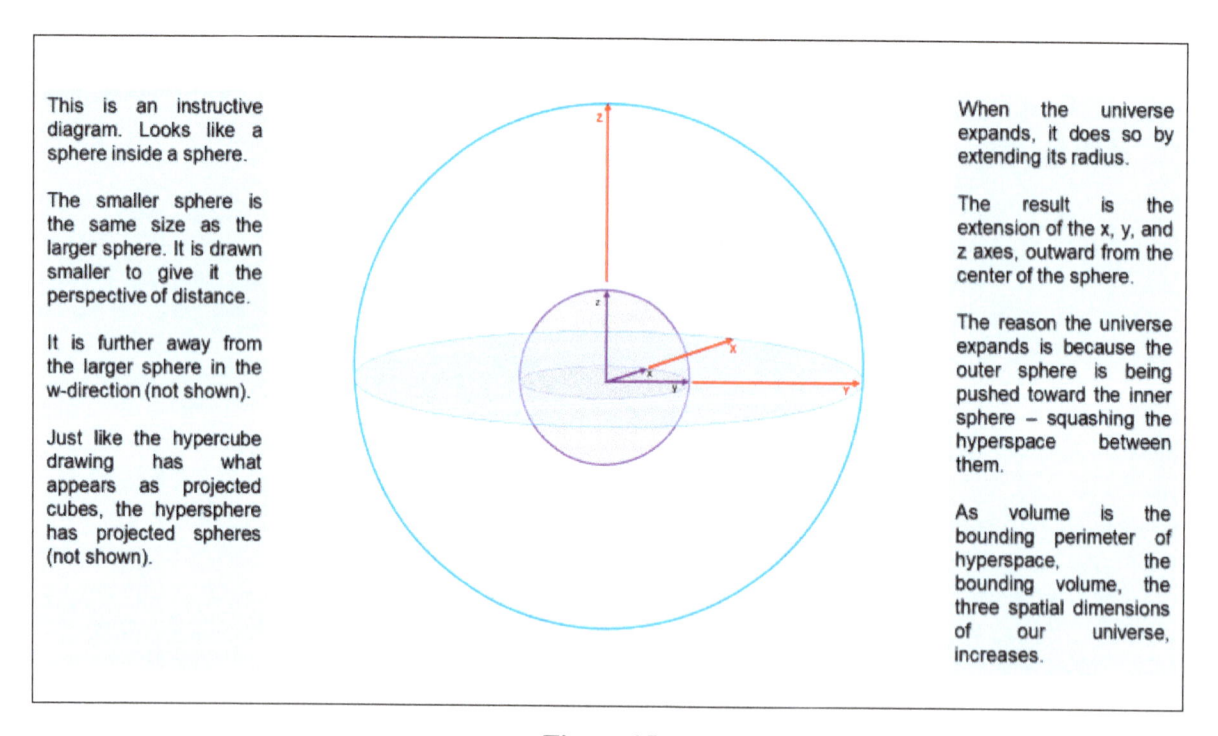

> This is an instructive diagram. Looks like a sphere inside a sphere.
>
> The smaller sphere is the same size as the larger sphere. It is drawn smaller to give it the perspective of distance.
>
> It is further away from the larger sphere in the w-direction (not shown).
>
> Just like the hypercube drawing has what appears as projected cubes, the hypersphere has projected spheres (not shown).
>
> When the universe expands, it does so by extending its radius.
>
> The result is the extension of the x, y, and z axes, outward from the center of the sphere.
>
> The reason the universe expands is because the outer sphere is being pushed toward the inner sphere – squashing the hyperspace between them.
>
> As volume is the bounding perimeter of hyperspace, the bounding volume, the three spatial dimensions of our universe, increases.

Figure 15.

It took us a while to get to this point. I hope you feel it was worth it. If I had jumped right in and told you that our universe was expanding because it is the boundary of a four-dimensional hyperspace being squashed – well, I would understand if you told me that you *just can't see it*.

And I do need you *to see it* – because there's more!

Of course, me saying that our universe is expanding because it's a volume perimeter being pushed in the w-direction and squashing a hyperspace – doesn't make it true. It's just an idea. But if the idea is rejected, then what might flow from that idea would be lost. Our task here is to either reserve judgment or take that idea as if it were true, and then see what implications might emerge when we give it serious consideration.

I thought about this for a while. I know you want proof. I wanted proof. Everyone wants proof! But the proof we seek does not come in the guise of a set of equations or experiments able to predict results. Rather, it just emerges. I know that doesn't sound very scientific – but believe me, that's what happens. Things emerge and fall into place in a way that pushes the proposition that the TRADD model we build here is a worthy contender. I put it forward here, as an alternative model for how the universe works. Why? Because as it accommodates each new perspective, it reveals to us, *why* it is that we make the observations we do. Things begin to make sense. Existing perspectives fail to explain, or poorly explain, the *how* and *why* of space, gravity, light, and time. The new TRADD perspectives, to me at least, appear to do a better job.

OK, so what does that really mean?

We have put forward a different perspective on space. It's not *nothing* – it is *something*. And because it is *something*, it can expand. That allows us to attribute the observed expansion of the universe to the expansion *of* space, even as we measure the expansion as the ever-increasing distances between objects *in* space. Later, we develop different perspectives on gravity, light, and time. We will find these new perspectives *rely* on space expanding in the manner described. If we describe the expansion of our universe in some other way, these new and different perspectives, including the new perspective on gravity, would elude us. What we discover within the TRADD context is that *how* and *why* space expands appears to underpin everything else. Our new perspective on space, *produces* gravity, *defines* the speed of light and *creates* the essence of time.

Now that's saying a lot. To understand all of it, we do need to take it one step at a time. We must persevere but start with the basics – the *mechanics* of our universe – spatial expansion.

Expansion in the x-w plane?

I want to make clear at the outset that we are using a hypersphere because for what follows, it is the most instructive. Our universe could be the bounding volume of a hyperspace having whatever shape one can imagine. It could be a hypercube, a hypersphere, or a hyperblob! It doesn't matter.

To show what is happening when space expands, we need to take 'a slice' of it – and examine that. We want to take a two-dimensional slice of our three-dimensional universe as it moves in the w-direction. But we are not going to take that slice in the x-y plane. Or the x-z or y-z planes. We want to see what happens in the x-direction when we move the sphere in the w-direction. To do that, we need to take a slice in the x-w plane. We can then see what happens in the x-direction when we move all of 'x' in the w-direction.

You see, if we have our large sphere in the x, y, and z directions, then when we move the sphere in the w-direction we expect our sphere to expand. And it will expand in the x, y, and z directions as the radius of the sphere increases. We can look at how the entire sphere expands but if we do that, it will be difficult to see what is happening in the w-direction – the direction in which we are pushing the sphere. So, to see what is happening, we take an x-w slice of the sphere that will show us how space expands in the x-direction when we move it in the w-direction.

Before we do that, though, just as before, we'll start with something familiar to help us orientate ourselves. We are all comfortable with the directions x, y, and z. So, let's begin our examination by looking at different ways to depict what happens in the space we're accustomed to. Let's take a slice in the flat horizontal x-y plane and move it up vertically in the z-direction. Then view what is happening if we only look at the vertical x-z plane as we move the horizontal x-y plane upward in the z-direction.

Don't panic if you got a bit lost there. With pictures and a little patience, you'll understand it all just fine.

Moving a sheet in the x-y plane in the z-direction

To do that, we take a square two-dimensional sheet in the x-y plane from the cross-section of the sphere, intersecting a square two-dimensional sheet in the x-z plane. We show the radius in each sheet (the red arrow in the z-direction and the red arrow in the y-direction), as this connects our understanding of how a sphere expands – a stretching of the radius.

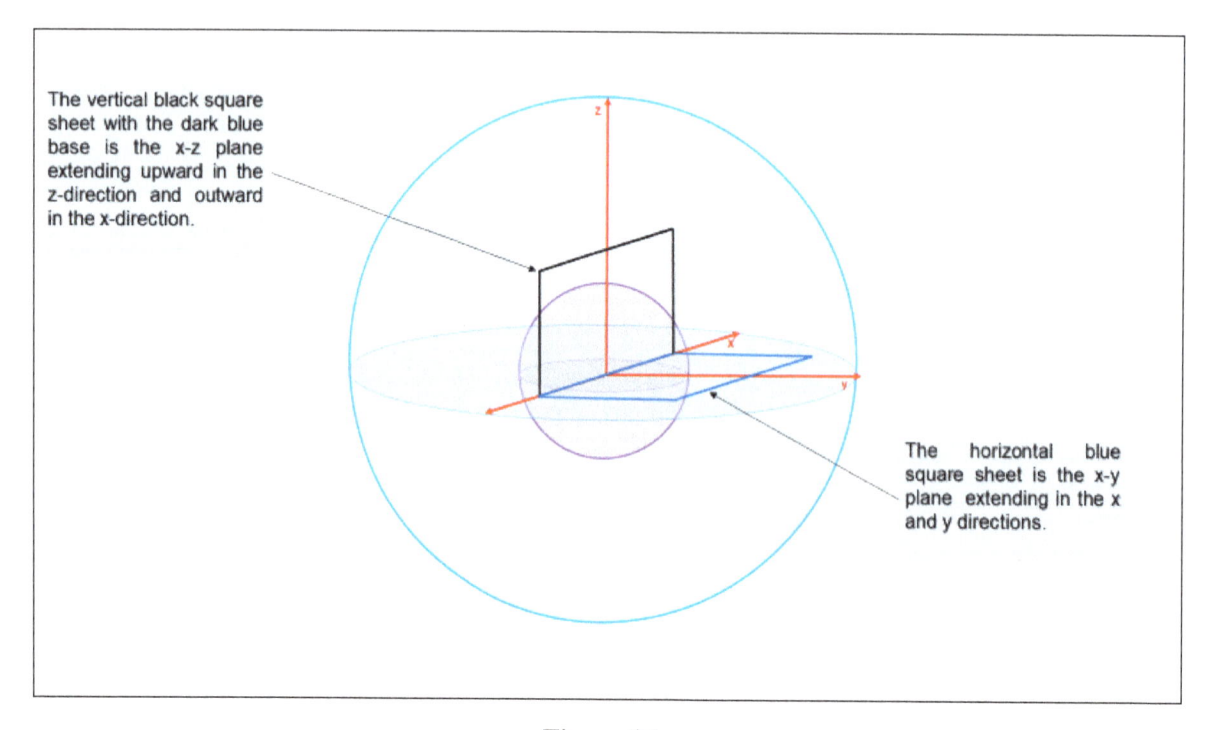

Figure 16.

What is space?

We then move the horizontal x-y sheet upward in the z-direction to show what that looks like.

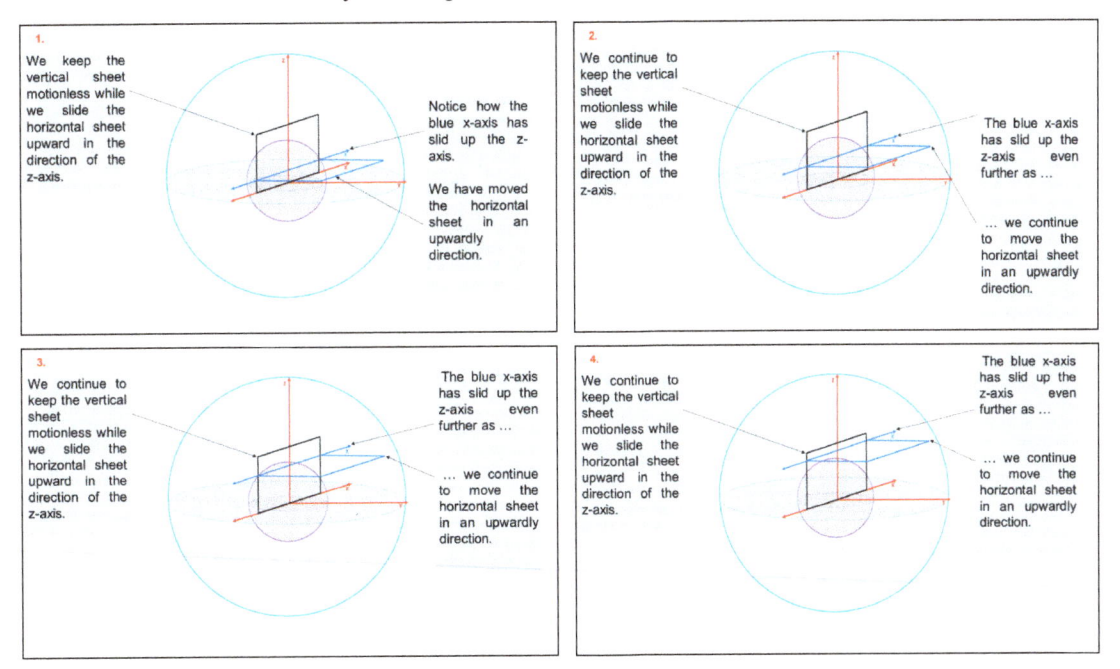

Figure 17.

Then we remove the y-dimension, so we are left with an x-axis only; a line moving upward in the x-z plane. The x-axis is where the horizontal x-y plane intersects the x-z plane as we move the x-y plane upward in the direction of z.

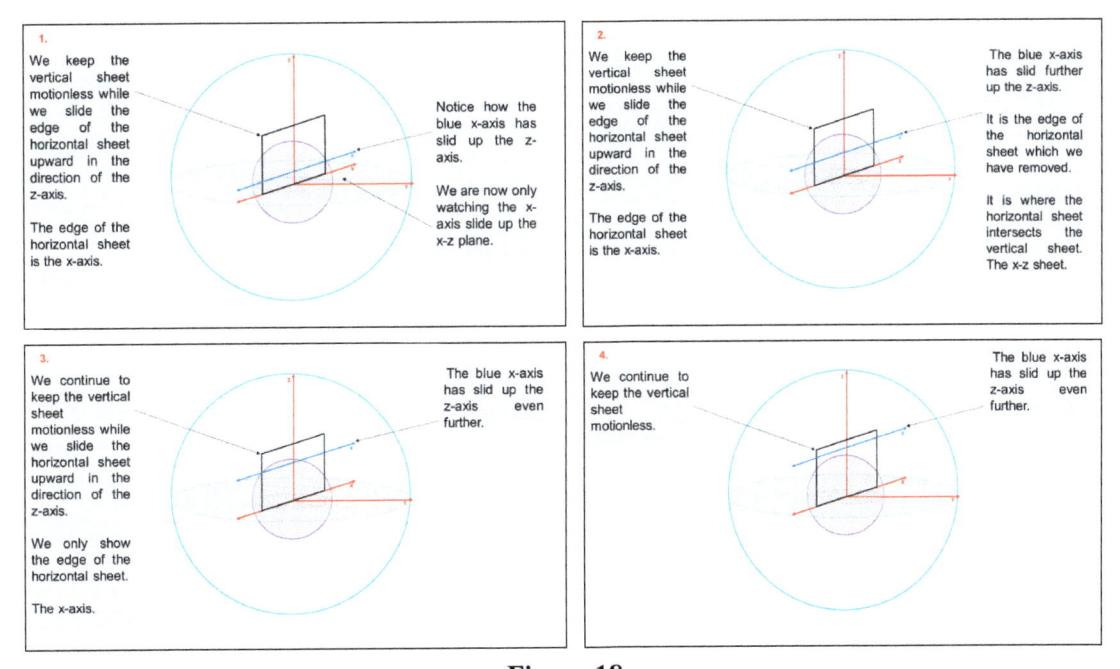

Figure 18.

Next, we want to look only at the vertical sheet in the x-z plane, and the blue x-axis, which is the edge of the horizontal sheet in the x-y plane. We want to move the x-y sheet upward in the z-direction as before. But this time, we will only show how the blue x-axis moves. The blue x-axis is where the horizontal sheet, which we are moving, intersects the vertical sheet.

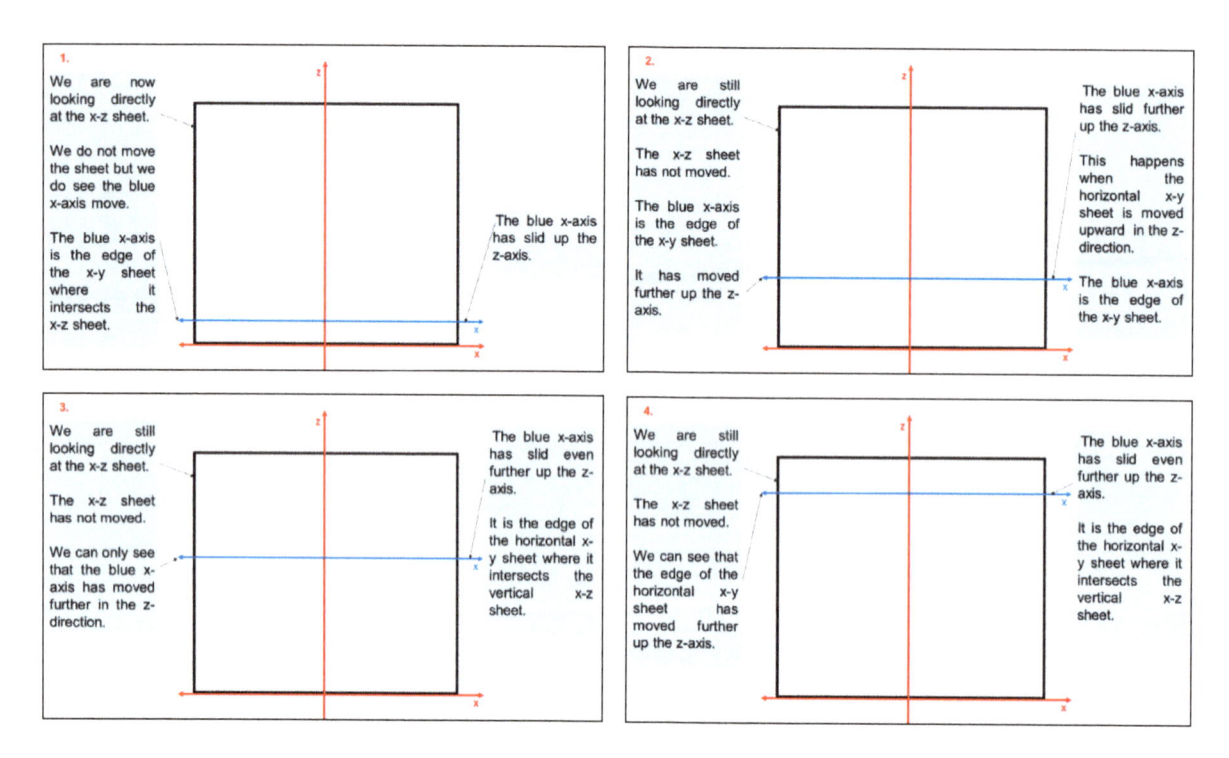

Figure 19.

Moving an x-y plane in the z-direction

Next, we want to remove the vertical and horizontal *sheets*. The vertical sheet lies in the x-z plane. The vertical sheet can be replaced by the entire vertical x-z plane. The horizontal sheet lies in the x-y plane. The horizontal sheet can be replaced by the entire x-y plane. What we are left with is that line where the vertical x-z plane intersects a horizontal x-y plane. That line is the x-axis. Then we want to move the entire x-y plane, up in the z-direction – as before. We can follow its progression by observing the upward movement of the x-axis.

1.
We are now looking directly at the x-z plane.

We have removed the vertical sheet.

The entire x-z plane is now the vertical sheet.

The z-axis does not move.

The blue x-axis has slid up the z-axis.

The x-axis is where the horizontal x-y plane intersects the vertical x-z plane.

We have removed the horizontal sheet.

2.
We are still looking directly at the x-z plane.

The z-axis has not moved.

The blue x-axis has slid further up the z-axis.

This happened because the entire x-y plane has slid up the z-axis.

The x-axis is where the x-y plane intersects the x-z plane.

3.
We are still looking directly at the x-z plane.

The z-axis has not moved.

The blue x-axis has slid further up the z-axis.

We removed the red x-axis.

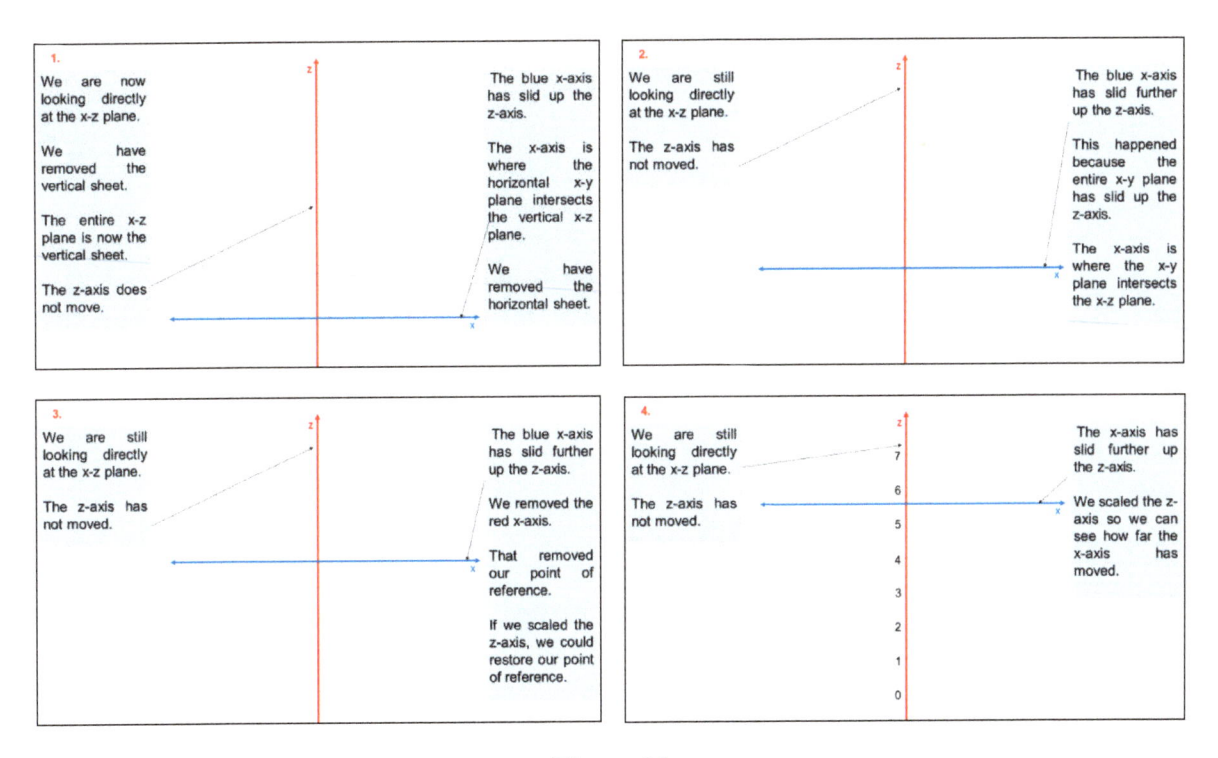

That removed our point of reference.

If we scaled the z-axis, we could restore our point of reference.

4.
We are still looking directly at the x-z plane.

The z-axis has not moved.

The x-axis has slid further up the z-axis.

We scaled the z-axis so we can see how far the x-axis has moved.

Figure 20.

Moving an x-y plane in the w-direction

Next, we replace the z-direction with the w-direction to create an x-w plane. Then we repeat the most recent exercise. We move an x-y plane intersecting the x-w plane, up in the w-direction.

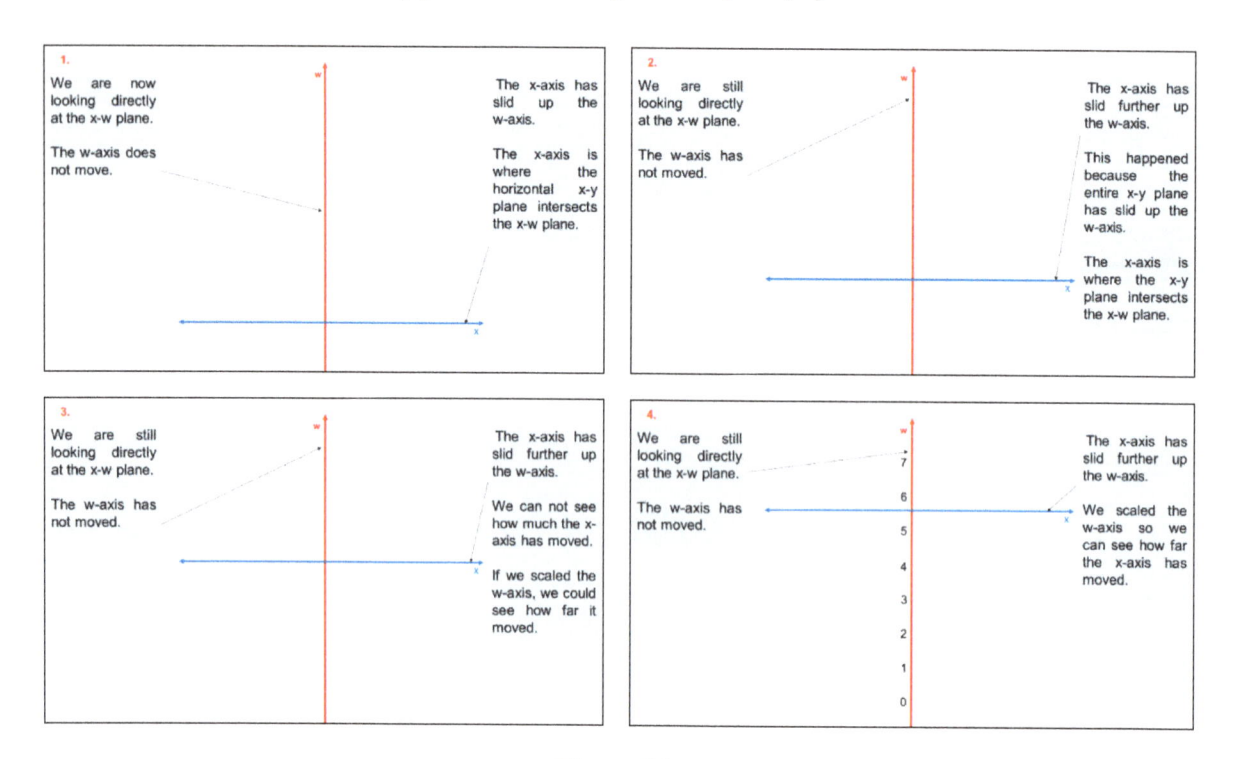

Figure 21.

So, let's review what we have done so far. We started out with a three-dimensional sphere that was part of the boundary perimeter of a four-dimensional hyper-object, a hypersphere. Our goal was to understand how it would look when we moved the sphere in the w-direction, squashing the hyperspace. For that to happen, the entire three-dimensional sphere would need to move in the w-direction. Because it's difficult to visualise how that would look in a way that we could take some measurements, we decided to break it all down.

First, we cut out horizontal and vertical sheets from cross-sections of the sphere. Then we took the horizontal sheet in the x-y plane and moved it up in the z-direction through the x-z plane. We showed how the edge of the horizontal sheet butted up against the vertical sheet in the x-z plane. We did not move the vertical sheet. We only moved the horizontal sheet upward in the z-direction.

We showed that the two sheets intersected along the x-axis. So, when we moved the horizontal sheet up the z-axis, the intersecting line, which was the x-axis, moved up as well because it was the edge of the horizontal sheet.

Finally, we removed the sheets altogether, leaving the x-axis to slide up the z-axis. We did this by looking at the x-z plane, face on, and moving the entire x-y plane upward in the z-direction. The entire x-y plane intersected the entire x-z plane along the x-axis, so we were able to follow the progress of the x-y plane in the z-direction by watching the progress of the x-axis.

Then we scaled the z-axis to restore reference points so that we could see how far up the z-axis the x-y plane had moved.

We did all *that* to show what it would look like if we kept one axis stationary (the z-axis) whilst moving an entire plane (the x-y plane) upward, along that axis.

When there are four directions, moving any one plane along the axis of a perpendicular plane will always produce the same behaviour. Because the w-direction is perpendicular to each of the x, y, and z directions, there are x-w, y-w, and z-w planes, and these are all perpendicular to the directions x, y, and z. That the x-z plane was perpendicular to the x-y plane was used as an example. The x-z plane is also perpendicular to the y-z plane. To understand this, think of the x-z plane as one side of a square box. The top and bottom of the box would lie in x-y planes. The front and back of the box would lie in y-z planes. So, both sides of the box would lie in x-z planes. All three variants of the planes are at right angles to one another.

Now, add a fourth direction: the w-direction.

We then replaced the z-axis with the w-axis, and we saw that the w-axis was a direction in the x-w plane. The w-axis is also a direction in the y-w plane and the z-w plane. In the first three dimensions, the z-direction in the x-z plane is also a direction in the y-z plane. Importantly, the x-z plane and the y-z plane are perpendicular to one another – they are at right angles. So, too, in four spatial dimensions, the x-w plane, y-w plane, and z-w plane all point in the w-direction and are all perpendicular to one another – they all meet at right angles. What that means is that when we move a sphere in the w-direction, we are concurrently moving the x-y plane, the x-z plane, and the y-z plane in the w-direction.

OK, you might be a little confused and that would be my fault. So don't panic. Think about a box again. When you move the front of the box by pushing it toward the back of the box, what you are really doing is moving a y-z plane in the x-direction. And when you move the left side of the box toward the right side of the box, you are really moving an x-z plane in the y-direction. Lastly, when you move the bottom of the box toward the top of the box, you are really moving an x-y plane in the z-direction – just as we did earlier as an exercise. Now that makes sense, doesn't it?

So, there are three planes when dealing with three dimensions. The x-y plane, the x-z plane, and the y-z plane. Each plane can move only in a direction other than the directions defining the plane. Hence, the x-y plane can only move in the z-direction, the x-z plane can only move in the y-direction and the y-z plane can only move in the x-direction.

But when there are four directions, *then* each plane has not one but two directions that do not define it. So, each plane can move two directions. The x-y plane can move in the z and w directions. The x-z plane can move in the y and w directions. The y-z plane can move in the x and w directions.

Oh! And when there are four directions, there are additional planes as well. They are the x-w, y-w, and z-w planes. And each of these can also move in two directions. The x-w plane can move in the y and z directions. The y-w plane can move in the x and z directions. And the z-w plane can move in the x and y directions.

So, the point is this. We can see now that the x-y, x-z, and y-z planes can each move in the w-direction. And, when that happens – when all three planes are moving all at once in the w-direction – we have a volume moving in the w-direction. The sphere is a volume. When we move the sphere in the w-direction, its three planes move in the w-direction simultaneously. So, we can understand how that looks by seeing what happens when one of its planes moves in the w-direction because what happens with that plane will also happen with each of the other two planes.

The x-w plane

We can now show the x-w plane, as a two-dimensional plane with a horizontal x-axis and a vertical w-axis. We are not interested in what is happening in the negative w-direction, so we only show an axis in the positive w-direction – the direction in which the x-axis is moving. Also, we need points of reference, so we scale both axes.

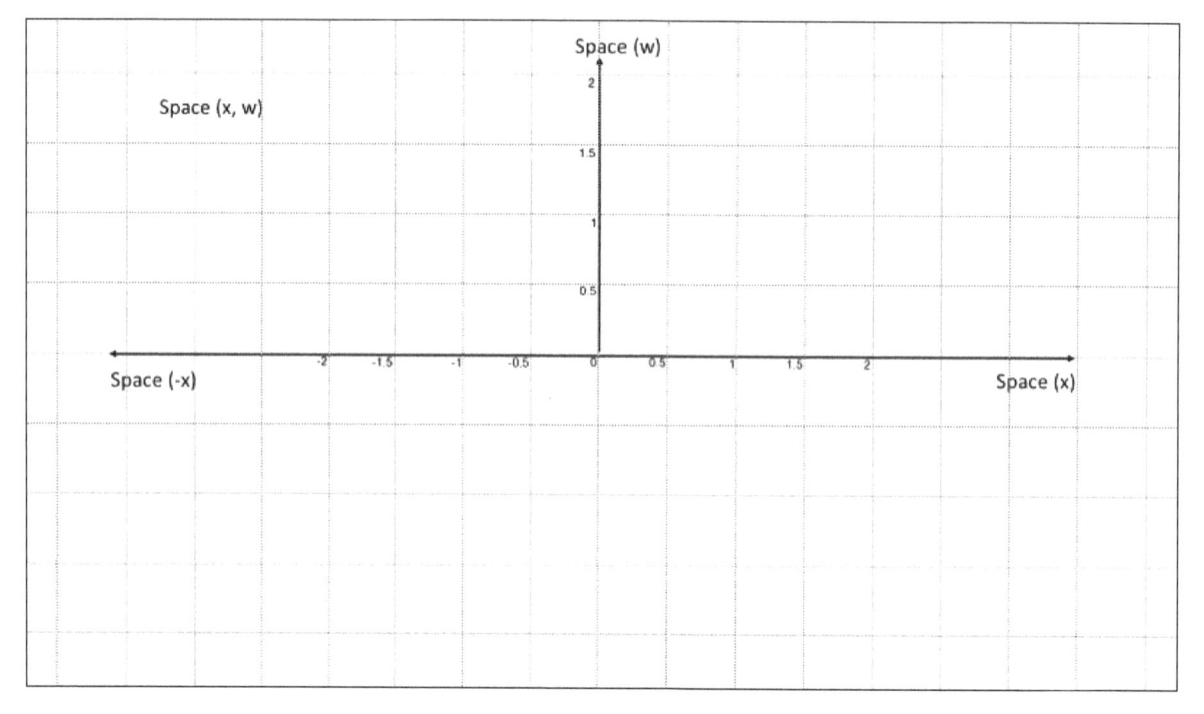

Figure 22.

Now we want to move the x-axis upward along the w-axis. We want to do this because we want to show in this x-w plane what it would look like if a sphere were moving in the w-direction. And because the sphere has an x-axis, the x-axis would need to move in the w-direction.

But we must consider that things do not move on their own accord.

Something must be pushing the sphere in the w-direction. And to push the sphere in the w-direction is to simultaneously push the x-axis, the y-axis, and the z-axis of the sphere in the w-direction. Therefore, something must be *pushing* on the x-axis.

And that's interesting!

We don't know *what* is doing the pushing or why.

But we can give whatever is *doing* the pushing a *name*.

Force.

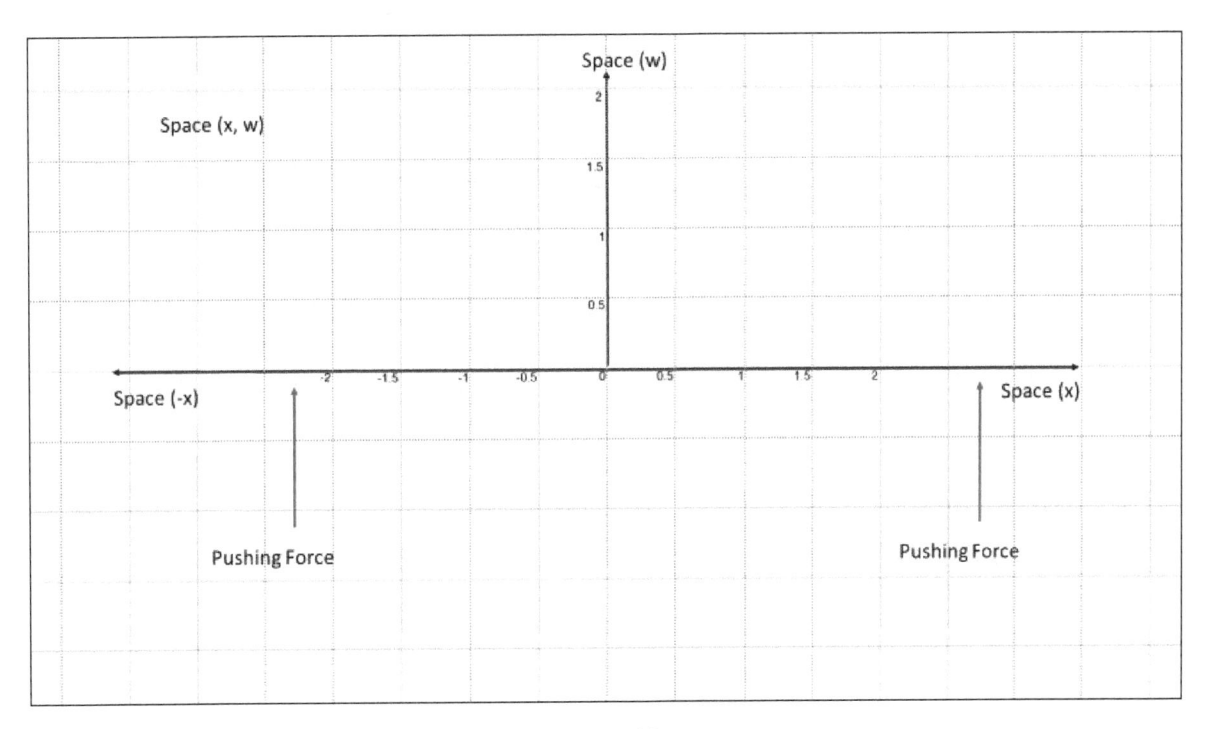

Figure 23.

Introducing forces

We can show that there is a force pushing the x-axis in the w-direction using purple arrows. There must also be a resistive force to the pushing force. Otherwise, the x-axis would move in the w-direction instantaneously. We know it doesn't and so we show the resistive force with blue arrows. The resistance to the pushing force.

The pushing force is pushing the x-axis in the w-direction. It is pushing from the x-axis to the w-direction so we can call it the [xw] force. The force resisting the pushing force is pushing back from the w-direction toward the x-axis and so we can call that the [wx] force.

Total Relativity and Dimensional Dynamics

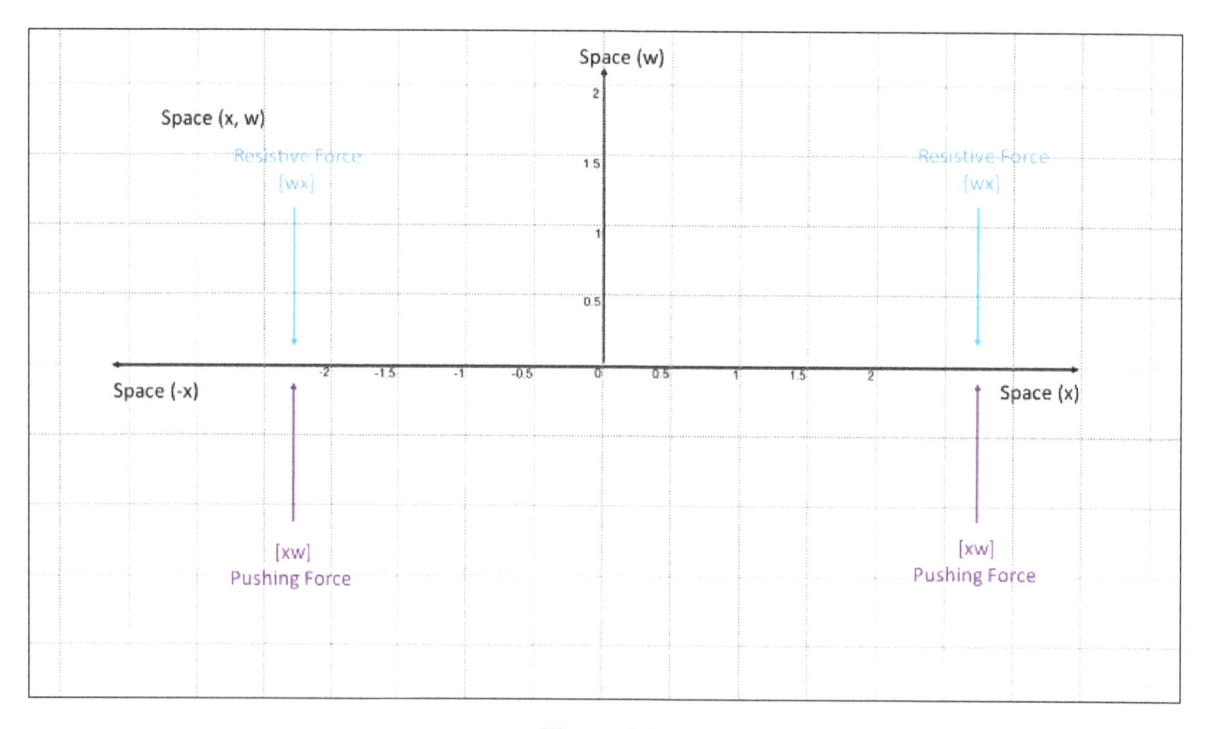

Figure 24.

As our sphere is pushed in the w-direction, it will expand. You may recall that the radius of the sphere has the same length in the x, y, and z directions. So, when the sphere expands, when its radius stretches, it will stretch evenly in all directions of x, y, and z. Therefore, we can show the expansion in the positive and negative x-directions on our x-w plane.

There must be a force responsible for stretching the x-axis and that force can only come from the w-direction or the negative w-direction. So, let's see what happens when the [wx] force appears to be responsible for stretching the x-axis as the [xw] force pushes our sphere in the w-direction. We can represent what is happening in our diagram.

As the x-axis is pushed in the w-direction, it stretches. The grey x-axis shows where it was, and the new black x-axis shows where it is after a duration of pushing. We can see the x-axis has stretched.

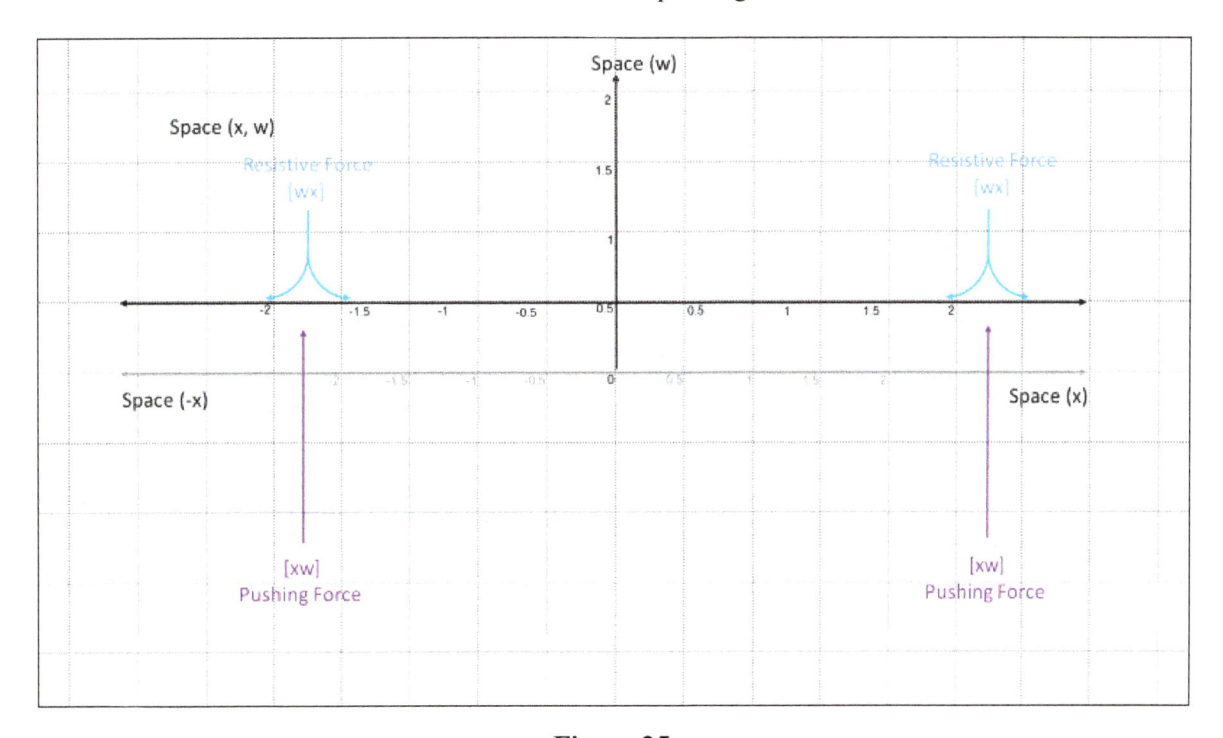

Figure 25.

Not only do we need to re-scale the x-axis, but we also need to recognize that there must be forces resisting the (successful) attempt to stretch the x-axis. Again, we reason that the resistance must exist to prevent the x-axis from expanding instantaneously. From the negative x-direction, we can call that force [-xw] and from the positive x-direction we can call that force [+xw]. And where [wx] divides, we can name the blue arrows and identify [w-x] opposing [-xw] and [w+x] opposing [+xw].

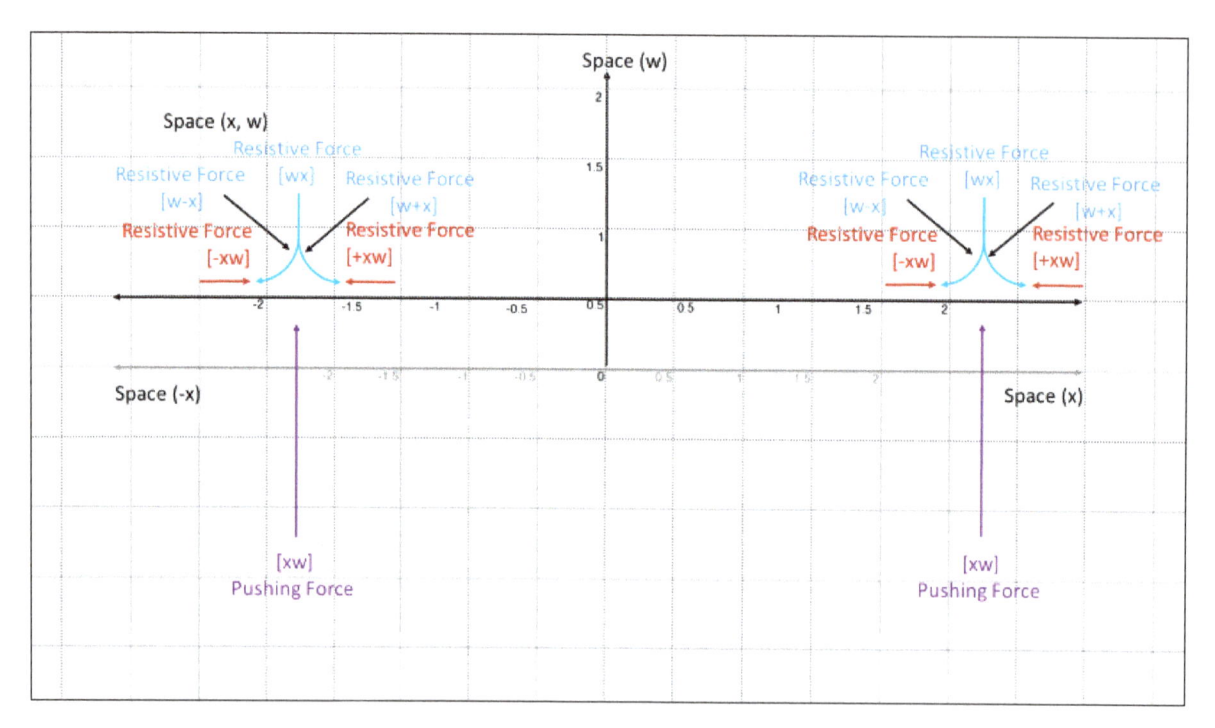

Figure 26.

Now, looking at our diagram, we can make some observations about the relative strengths of the forces we have identified.

[xw] is stronger than [wx]. Well, it must be, because it's winning.

After losing its battle with [xw], [wx] divides its strength to stretch the x-axis. And interestingly, this is a battle it does win. So, we know that the combined strength of [-xw] and [+xw] is not powerful enough to beat [wx] but [xw] is. That tells us that [-xw] + [+xw] is weaker than [wx] and therefore weaker than [xw] (which is more powerful than [wx]). Fun, isn't it?

Relative strengths

So, we can summarize in order of strength – strongest at the top of the list.

[xw]

[wx]

[w-x], [w+x]

[-xw], [+xw]

But we know that this occurs in the y-w and z-w planes as well. So, we can write.

[xw], [yw], [zw]

[wx], [wy], [wz]

[w-x], [w+x], [w-y], [w+y], [w-z], [w+z]

[-xw], [+xw], [-yw], [+yw], [-zw], [+zw]

Now, what if there's a relationship between these forces and the forces in the standard model? We know, for instance, that the strong force experienced by quarks, gets stronger the further two quarks move away from one another, like the tension in a rubber band as it gets stretched. The further you stretch it, the stronger the force pulling against it being stretched. But isn't this exactly what is occurring along the x-axis? It is the [-xw] and the [+xw] forces that oppose the x-axis stretching. To constrain its expansion to a rate.

There are three directional axes in our universe, and three very strong pairs of forces acting against those forces acting to stretch these axes. As our universe expands, it does so at a rate. That rate is governed by the forces resisting the expansion. They are [-xw], [+xw], [-yw], [+yw], [-zw], [+zw]. Could they present to us as quarks? We have talked about an n-dimensional object being pressurised so that a deformation at its boundary results in the boundary expanding. The internal pressure of the object presents at the boundary as a resistive force – resisting the deformation. Presenting as a force to all three axes in this way we have [w-x], [w+x], [w-y], [w+y], [w-z], [w+z]. In the standard model, we have gluons. Could gluons be the physical representation of the forces pushing against our universe from the w-direction?

I'll leave these questions for others to ponder. Toward the end of our study of the TRADD model, we will be left with a sense of its plausibility or not. And if plausible, then surely the forces identified here will need to be mapped onto the forces identified under the standard model. Recall that the TRADD model presents as an alternative perspective on what's going on. It does not seek to replace the standard model. Hence, if the perspectives drawn out in the TRADD model survive the plausibility test, then as both models are equivalent but from different perspectives, the TRADD forces ought to map onto forces in the standard model.

[xw], [yw], [zw]

[wx], [wy], [wz]

[w-x], [w+x], [w-y], [w+y], [w-z], [w+z]   (Are these associated with gluons and the strong force?)

[-xw], [+xw], [-yw], [+yw], [-zw], [+zw]   (Are these associated with quarks and the strong force?)

Force requires energy to act. Mass can be thought of as a volume of space where energy is concentrated. Energy and mass are said to be different manifestations of the same thing. Perhaps we can think about mass as condensed energy. So where forces are experienced by quarks, say, that is not to say that those forces only act where there are quarks. TRADD shows forces acting everywhere in space. Perhaps they are just more easily identified when concentrated in matter. After all, we don't measure space – only things in space.

This is, of course, purely speculative at this juncture. The rationale for the forces is that, firstly, something must be doing the pushing – implying a force. But if there were no resistance to that force, then one must ask what governs the pace with which the x-axis moves in the w-direction. If it were not governed, then one could imagine that this process occurs instantaneously.

We know from our thinking about dimensional dynamics that squashing a spatial dimension whilst conserving it results in an expansion of its bounding spatial dimension. And as we have seen from our diagram, that expansion takes place in the positive and negative directions of the x-axis. Any 'expansion' in the w-direction doesn't make sense because the w-direction is not one of the three directions defining the volume bounding the hyperspace of the hypersphere.

So, we are compelled to stretch the x-axis but are again challenged to ask if that stretching happens in an instant or at a rate. Clearly, if the progression of the x-axis in the w-direction is measured (by 'measured' I mean that it is happening at some rate) then the stretching in the x-axis must consequently, also be measured, at a different, perhaps related rate.

Now, that's interesting!

A *rate* implies time.

But what is it?

What is time?

Introducing time

When we scale the x-axis, we can give it units in metres. We could scale the w-axis and give it units in metres, too. But whilst we can travel in the x-direction and measure metres off using a ruler, we can't do that with the w-direction. We can't measure metres off using a ruler in the w-direction. We can't travel in the w-direction.

Or can we?

We think of time as something that passes, and we have given it units in seconds. We could say 'time passes' or that we are – moving *through* time. In our model, the x-axis moves *through* the w-axis. Or, in a relativistic sense, were we to regard the x-axis as stationary, then we would describe that movement as 'W' passing. When we drive in our car, we can think about driving through the countryside or we can look out the window and see the countryside pass us by. It's a matter of perspective. If we regard ourselves as moving, then we *move* through the countryside. If we sit in our car and regard ourselves as stationary, then as we look out the window, we see the countryside pass by.

Can you see where this line of thought is going?

Creating time

Let's first give the w-axis units in metres. We also give the x, y, and z axes units in metres. We do this initially, because we are dealing with space in four dimensions – in four different directions – x, y, z, and w. Now, in our model, we don't have the x-axis moving through the y-axis or the z-axis. In fact, the x, y, and z axes are stationary with respect to one another. But all three x, y, and z axes – the diameter of the sphere in length – move through the w-axis. The entire sphere moves in the w-direction but the sphere itself has only directions in the dimensions of x, y, and z.

Renaming an axis does not change what it is. When we put numbers on it, an equal distance apart, that's all we do. The distance between the numbers on the w-axis is the same as on the x-axis, the y-axis, and the z-axis. So, they each have the same scale – at least regarding the way we have numbered them. Giving the numbers units, such as metres or seconds – that does not change the distance between points on the graph. A point two units up the w-axis and two units along the x-axis is at the coordinates (2, 2). It does not matter what you call the units in the w-axis or the units in the x-axis.

So, let's do two things.

1. Let's change the units in the w-axis from metres to seconds.
2. Let's change the name of the axis from the w-axis to the time-axis.

Now, our x-axis is no longer moving up the w-axis but up the time-axis. We no longer measure how far it has moved in metres but in seconds.

Can you see that giving something a different name and units doesn't change what it is?

A good example is when we compare distances measured in feet and inches to distances measured in metres and centimetres. When we measure it in feet and inches, we say it is the distance measured in imperial units. When we measure it in metres and centimetres, we say it is the distance measured in metric units. And if we measured it in hours and seconds, we *could* say it is the *distance* measured in time units.

Just as an inch is not the same distance as a centimetre, an inch is not the same distance as a second. But a second will be a *distance* none the less. The only difference is that in our everyday world, we don't think of ourselves as being able to travel any distance in the w-direction. In other words, if we think of the w-direction as distance, then we believe we *can't* traverse it. But if we think of the w-direction as time, then we believe we *can* traverse it. Or at the very least, we say that time *passes* – in other words – *we travel through it*.

Trick or treat?

Now I know what you're thinking. That this is 'some kind of trick' I'm playing on you. But let me reassure you, it's not. All we did was change the units from metres to seconds and called it time – that's all we did. It is critical to understand that we did not change what it *was*. We only changed how we describe and measure it. And by 'it', I mean the w-direction.

I can imagine that even this simple adjustment – calling the w-direction the time-direction and changing its units from meters to seconds whilst insisting we are still talking about distance – well … that will raise some eyebrows. After all, time and distance are different things, right?

Well … not really.

Let me explain.

If you look up what time is, you will get something like this. A second is the time that elapses for a specific number of cycles of the radiation produced by the electron transition between two shell levels of the caesium 133 atom. Yea, I tried to simplify the definition but it's still a mouthful. So let me break it down for you. Radiation produced by the transition of an electron between two shell levels of the caesium 133 atom is an event. When a specified number of these events have occurred, one second has elapsed. Put another way, it takes one second for a specified number of these events to occur.

Ok, but what does that really mean? It means that if we count these events as they occur, then when the count gets to that specified number, one second has elapsed. And if we continue to count, then when the count gets to twice that specified number, then two seconds have elapsed.

Travelling through time

What it does *not* say is that whilst we are sitting there counting away, the x-axis is moving up through the time-axis. Let's say that the x-axis moves up the time-axis at a constant speed. If you were at a point on the x-axis, you would be moving up the time-axis at a constant speed (and probably measuring that off in seconds). We don't think of ourselves moving in the time-direction in the same way as we think of ourselves moving in the x-direction. But what if we did?

Let's jump into a car and travel at a constant speed, say 60 km per hour, along the x-axis. Remember, we *can* travel in the x-direction. Now, whilst driving along we can count our caesium 133 atom events. We will find that for every metre we travel, the count will be the same. In other words, there will be a count value, (let's call it Blah) that will be reached after we have travelled one metre when travelling at the constant speed of 60 km per hour. And after travelling two metres, the count will be two times Blah. After travelling for an hour, we will have travelled 60 km or 60,000 metres and if we had continued counting, the count would be 60,000 times Blah. In other words, 60,000 times the count value of Blah, is the number of caesium 133 atom events that will have taken place after travelling 60 km.

You see, we *can* use the counting of events to measure distance, so long as we travel over that distance at a constant speed – a constant rate. When travelling at a constant speed, whether we say we have travelled for an hour or travelled 60 km is simply a matter of units.

If we start at a point A on the x-axis and end up at the point B, we can call the distance between A and B, 60 km or we can call it 1 hour. The distance is the same. But we can choose to measure it in units of time or units of length. If we start at A and do the count, we will end up at B every time. It's just that when moving in the w-direction, when we end up at B, we say the distance travelled was an hour. And when we travel in the x-direction, when we end up at B, we say the distance travelled was 60 km.

Measuring time and distance

Now here's the 'kicker', as they say. Once we have established the relationship between km and hours, we can stop counting events. We use a mechanism to measure the kilometres travelled: the odometer in the car. We use a mechanism to measure the number of hours we have travelled: the watch on our wrist. Then we can use either mechanism to know the value reached on the other. When the odometer reads 60 km, we know the watch will read 1 hour. And when the watch reads two hours, we know the odometer will read 120 km. Our counting of events enables us to calibrate the relationship between different units used to measure our travel progress. Once the relationship is established, we can stop counting. If I have a mechanism to measure miles and another to measure km, then I can count events until I have travelled one km and look at the value of the mechanism measuring miles. I find that 0.625 miles is the same as one km. After that, I no longer need to count events. I can look at my odometer and read off 60 km and know that when I look at the mechanism measuring miles it will read 37.5 miles.

We already do this.

Another definition for time is how much of it has elapsed after light has travelled a particular distance. We can do this because after taking measurements we find that light always travels at the same speed and at a constant speed. So, we know the relationship between time and distance. We have checked that relationship by counting events. But once the relationship has been calibrated, then we only need to know one of the units to derive the other. If we know how far light has travelled, then we know for how many seconds it has been travelling. If we know for how many seconds light has been travelling, then we know how far it has travelled. It is the calibrated relationship between the different units we ascribe to travel, that enables us to infer the value of one unit from the other. But the way we do the calibration is by counting events.

So, what am I saying?

When travelling at a constant speed of 60 km per hour, 60 km east then 60 km north is no different from 60 km x-direction then 60 km w-direction, which in turn is no different from 60 km x-direction then 1 hour w-direction which is no different from 60 km x-direction then 1 hour time-direction.

Renaming the w-axis

We have introduced the idea of time into our model. Time is the distance travelled by the x-axis in the w-direction. So, in our world, the TRADD world, 1 *year* represents a distance travelled by a point on the x-axis at a coordinate defined by its position in x, y, and z … in the w-direction.

If we scale the w-axis in *years* and call it the time-axis, then time is the number of years travelled by the x-axis in the time-direction.

One light-year is the distance in metres that light travels in 1 year. As a *speed*, it is also referred to as a universal constant. We can mark off the x-axis in units of light-years and the time-axis in years.

Total Relativity and Dimensional Dynamics

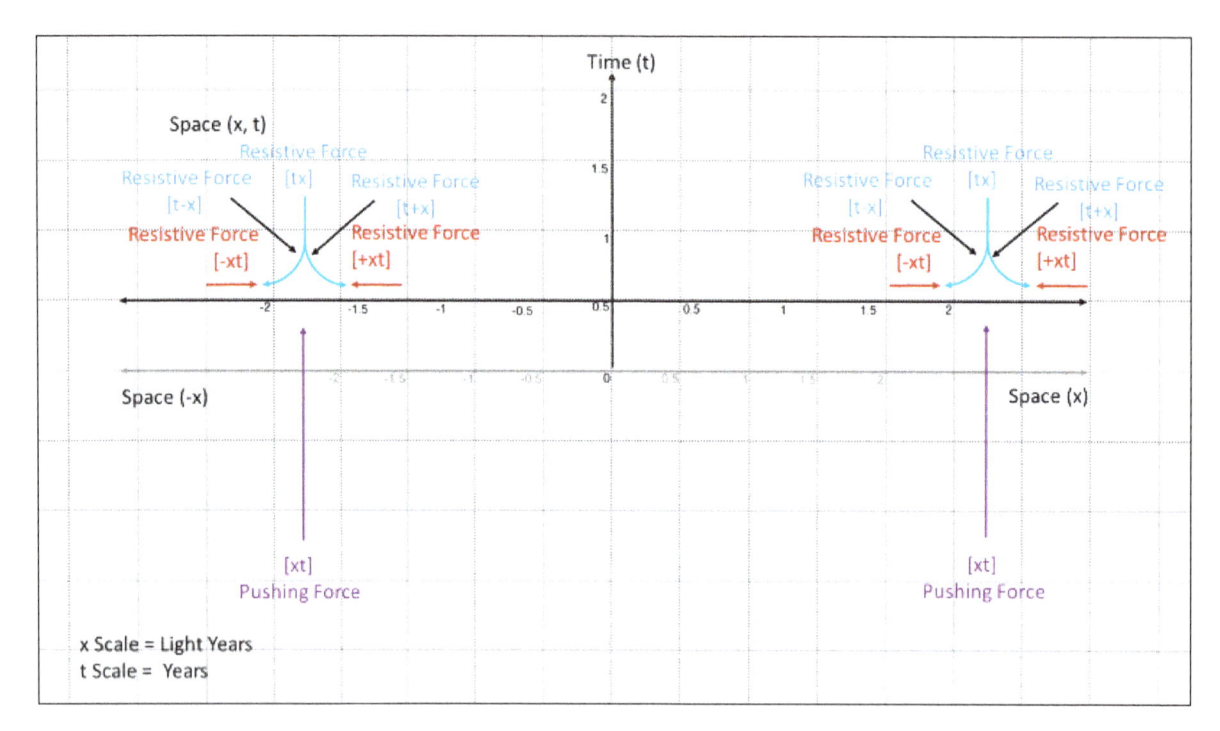

Figure 27.

When we do that, however, we need to rename the identified forces. We simply replace the letter *w* in the name with the letter *t*.

It would be interesting, having introduced time, to continue exploring these forces.

But we have bigger fish to fry.

Because now, we need to deal with the *beast*, gravity.

What is gravity?

Gravity is a beast to explain and understand. But is it a force? To understand *that*, we need to take a *really* close look at what's going on in space. We also need to study what the standard model has already discovered. We need to get a sense of what Einstein figured out to give us general relativity and how his field equations for gravity perfectly describe *one* perspective of what's going on. Then, we will see how a different perspective clears up some of the mysteries surrounding Einstein's new account of gravity.

So, having modelled 'space' (and spacetime … recall the w-direction), we can now turn our attention to the beast, to 'gravity'.

What is it?

The standard model

The standard model tells us it is a force that supposedly acts over any distance (out to infinity) and acts to attract objects having mass, toward one another. This force is said to manifest in mass as a property of it. An interpretation is that an object having mass exerts the force called gravity, which bends spacetime in a way that makes other objects having mass *fall* toward it.

The amount of gravity an object has is proportional to its mass. I hasten to add that when we talk about mass in this context, we must include energy because it's been shown that mass and energy are interchangeable (or for the purists out there – mass and energy are equivalent – they are different manifestations of the same thing).

Hence, the standard model is telling us that the force we named gravity manifests in mass and energy and that *its power* is to bend spacetime. I say '*its power*' because that's how gravitational force works. It doesn't *really* attract masses to one another – although that's the observed effect. Gravity *bends* spacetime, that's all, that's it – that's what it does. Because of spacetime being bent, out of shape as it were, things happen. One of the things that happens is that objects having mass *fall* through spacetime toward other objects having mass. Two objects having mass will therefore fall toward one another. We observe that behaviour but interpret it as objects having mass being attracted to one another by the force we call gravity.

And the further away from an object having mass, the weaker the gravitational 'force'. So, spacetime is bent *a lot* close to the object, but the further away you get, the less spacetime is bent. What (to me anyway) is questionable about this interpretation is that an object having mass can bend spacetime, if ever so slightly, over any distance out to infinity. Hence, an atom (which has mass), here on Earth, can bend spacetime, on the other side of the universe. Hmmm.

I'm not trying to trivialise the force of gravity here. What I *am* trying to do is to be absolutely crystal clear about what the standard model *says this force is* – or more specifically, *what it does*. It bends spacetime.

Another idea, long believed, is that space is *nothing*. And even with the invention of spacetime in the standard model, that, too, is *nothing*. There was a time (going all the way back to the Greek philosopher Aristotle), that empty space was thought to be *something* – called the *ether* – and at the time, a requirement for the propagation of light, which was thought not able to propagate through empty space – through *nothing*. But today, in the standard model, spacetime is indeed just that – *nothing*.

Now you do have to wonder.

On one hand, we are told that spacetime is *nothing*. On the other hand, we are told that gravity *bends* spacetime. Gravity bends *nothing*! And after taking *nothing* and bending it, *something* – an object having mass – moves through spacetime in a different way. It *falls* toward where the spacetime is *bent*. A *force* bends *nothing* so that *something* can *react* to *that part* of the *nothing* that was *bent*.

Really?

And then there's the 'elephant in the room', which no one, best as I can tell, is talking about. Let's take a body, like the Earth, revolving around the Sun as it is. As it moves through a location in spacetime, the spacetime in that location gets warped by the Earth's mass. But then as Earth moves on, that warped spacetime snaps back into place. The gravitational field moves with the Earth as the Earth moves away from that location. So, if we say that mass bends spacetime, then what *unbends* spacetime? Why does it not stay bent? It suggests that even as mass exerts a force on spacetime to bend it, the fact that spacetime unbends – snaps back into place – suggests a resistance to the force doing the bending. We know mass is equivalent to energy and hence able to fashion a force and exert it on spacetime. But for spacetime to snap back, suggests it is resisting that bending for the duration it is bent. In other words, it suggests that somehow spacetime can exert a force, which requires energy. That suggests spacetime has energy. It also suggests that the shape preferred by spacetime is a flat shape – not the curved shape produced in the presence of mass. Go on. Tell me you don't find this just a little bit curious. Space snaps back into place. That's extraordinary, surely. It's possibly more significant than mass bending spacetime in the first place! But – show me where it even gets a mention. Unbelievable!

Maybe I've misinterpreted something. Maybe I'm creating drama out of nothing. But maybe, that's just what we need. A little bit of drama, so that we might begin to think about things in a different way. For me, the idea that space is nothing and then, well, we'll just bend it – absurd.

Can you feel my frustration? My belief is that people doing science feel that frustration, too. But our predictions so precisely match our experimental results – even when we repeat those experiments many times. We have a *theory* in our standard model. That *theory* makes *predictions* about how the physical world ought to behave. We set up experiments to test that *behaviour*. Our experiments *produce* the predicted behaviour. And so, *the theory must be true*. It's difficult to argue against that, isn't it? But what I am attempting to do here is not so much argue against it as to suggest that the way we *interpret* what we *observe* – well, that could use an upgrade.

Bent straight lines

When I first began to think about gravity, of particular interest was the effect observed, where the path light takes, will bend as it passes close to a massive object in space. The standard model explains that the mass of the object bends spacetime. If one were to draw a straight line close to but passing by a massive object, the line would bend. Imagine a straight line drawn on a piece of paper. If you bend that piece of paper so that it no longer lies flat, say, lift one corner, what happens is that the straight line drawn on the paper would bend, too. It bends because the paper is bent. So it is, with straight lines in spacetime. They are straight lines through spacetime, but where spacetime is bent, the straight lines bend, too.

The standard model also says that light travels in a straight line. Well, that's until the straight-line path it's travelling on becomes bent due to a bending of spacetime, due to … yes, you guessed it … due to gravity.

So, light travels through spacetime along a path that is a straight line that curves wherever spacetime is curved due to gravity.

One might conclude that since gravity acts over an infinite distance, that it bends spacetime everywhere, and hence, there does not exist a straight line through spacetime anywhere in the universe – except by chance. For example, where spacetime that's already bent, is bent again so that it ends up straight.

According to the standard model, that's how gravity works to effect spacetime and that's how light moves through spacetime – in a straight line that's either straight or bent – but mostly bent.

Now *that* … is fascinating.

Newtonian gravity

One of the things I struggled with was the idea that gravity was a force acting over a distance – any distance.

I can understand the genesis of this idea from the time of Sir Isaac Newton. At that time, there was no spacetime continuum. In Newton's time, there was neither special relativity nor general relativity. For Newton, the idea that mass caused spacetime to bend was not a consideration. Hence, his idea of an invisible attractive force generated by an object having mass, pulling on other objects having mass made sense – particularly as physical experiments demonstrated that this interpretation explained what was observed – well, most of the time. Newton also explained that the distance separating two objects was significant, so that as the distance increased, the attractive force weakened. Hence, also, as the separating distance decreased, the attractive force strengthened.

Newton also showed that the amount of mass an object had was proportional, in part, to the magnitude of the attractive force. He showed that it was the product of the masses of two attracting objects that determined the size of the attractive force between them, before it was weakened by the separating distance. As the distance decreased, the weakening of the force decreased – so that the force became stronger and stronger. As the distance increased, the weakening of the force increased – so that the force became weaker and weaker.

In this scenario, the mass of each of the two objects remained constant, whereas their separating distance changed over time. As objects attracting one another moved closer together – they would eventually collide. Moreover, because as the distance decreased, the weakening of the force decreased – so that the force became stronger and stronger – the speed with which objects travelled toward one another due to the attractive force increased – the objects accelerated!

He also showed how a less massive object travelling with some velocity toward but destined to pass a significantly more massive one, would similarly be affected by this attractive force to the extent that its direction of travel would be changed by it – the travelling object's trajectory would be altered due to the force of gravity.

So, we are presented with a challenge here. To arrive at a result where the force of gravity *does not* act over a distance – recalling that we have measurements that prove it does – we essentially need to come up with a different explanation for what is measured. We need a *different way* to interpret what is observed, and indeed experienced. To do away with a force able to act over a distance, or a force able to bend spacetime, we essentially need to do away with the force itself.

Getting rid of gravity

First, we must conclude that gravity is not a force. It's interesting that some interpretations today talk about gravity as an *apparent* force. It's as though the whole idea of gravity bending spacetime has resonated with some of the thinkers out there, to the point where they recognize how general relativity, in its new description of gravity, somehow sits at odds with the Newtonian idea of gravity being an invisible attractive force between two bodies. It makes sense when you think about it. We can't on one hand say that gravity is a force that attracts two bodies having mass (Newtonian), and on the other hand say that gravity is a force that bends spacetime and that it is the *bent spacetime* influencing the direction in which mass travels that results in what we measure as an attractive force between the two bodies having mass (general relativity).

Second, we need to figure out what's really going on. What, if not gravity (bent spacetime), produces the observed behaviour that we can measure and experience?

We essentially need to do away with gravity all together – so it doesn't exist. It can't. Because if it did, then it would be gravity as we know it. It would be a force (or an apparent force). And that force *can* act at a distance. And that's something that we want to utterly reject!

So, we want to get rid of gravity.

Then, we want to reinvent it.

Reinventing gravity

We got rid of gravity. We now need to reinvent it. After we do that, we can still call it gravity, but our understanding of what it is will be different. It won't be a spacetime bending force. It won't act over a distance. But all the calculations we do now, our equations, must still work. Recall that we do not seek to debunk the standard model. Rather, we seek to provide a clearer understanding of what the standard model is telling us.

Now if that sounds like a bit of a contradiction, well ... you're right – it is. On one hand, we are saying that gravity does not exist. On the other hand, we are saying that, yes, having reinvented it – it does. So let me explain.

The reason we got rid of gravity is because in the standard model it is a spacetime bending force. More significantly, it is a force able to act over a distance. The idea that a force could act over a distance is something we reject. We therefore reject the idea that gravity is *that* force. We start out with the idea that gravity is *not* that force, and thereby reject it – completely.

But we accept that there are observations that, when measured, fit neatly into the standard model labelled as gravity. And in the standard model, calculations using equations involving gravity always seem to come out right. So, there *is something* going on.

We can't get rid of gravity and then not acknowledge what is observed and measured. Something is going on that produces behaviour that we can reliably measure consistently. Now when we figure out *what* is going on, what should we call that behaviour? The behaviour is already dealt with in the standard model as due to gravity. Producing a different *interpretation* for what is going on won't change how that behaviour is accounted for in the standard model. What we measure, put into equations, and calculate with won't change. Only our understanding of what we are measuring will change. So, it makes sense to persist with the name 'gravity'. After all, when Einstein changed the description of gravity from an 'attractive force' to a 'bending of spacetime', *he* still called it 'gravity'.

And therein lies our first clue.

How Einstein killed gravity

Gravity in the standard model already got an upgrade when general relativity described it as a bending of spacetime. Before general relativity, gravity was an invisible attractive force acting over a distance. Newtonian physics if you like. Complete with equations. The new description – that of spacetime being bent – brought new equations. But the *description itself* already dispels the idea that gravity is an attractive force acting over a distance. Recall I noted that some descriptions now already describe gravity as an *apparent* force.

Let me be clear.

The *equations* still represent gravity as a force – and as a force acting over a distance. The new *description,* however, barely describes gravity as a force, much less one acting over a distance ... perhaps.

When you consider the new description of how gravity works, you are left simply with – that matter (having mass) bends spacetime. Oh, that's the first of a two-part description. The second part of the description is that matter will move through spacetime according to how spacetime is shaped, (according to how it is bent.)

John Wheeler, an American theoretical physicist, galvanized this perspective in the book *Gravitation*[2], saying, 'Mass acts on spacetime, telling it how to curve. Spacetime acts on mass, telling it how to move'. One popular retelling of this is that 'mass tells **space**time how to curve, and **space**time tells mass how to move'. Notice the emphasis is on space. It's like we still don't fully understand what spacetime *is*, so we draw back from that to the familiarity of space because it's something we are all comfortable with.

Now, you really have to wonder – don't you?

Here, we have *two* descriptions required to make sense of gravity in the current (2023) standard model. Putting time aside momentarily, we get: Matter bends space. How space is bent determines how matter moves. It's a thorough Yin-Yang composite!

But how, then, is gravity a force?

It used to be – *when we said it was* – when our description of it from Newton was some invisible attractive force generated by the presence of matter. The more matter (the more mass, actually), the greater the force. The closer you were to that matter, the stronger you felt the force. The further away you were from that matter, the less you felt the force.

Gravity *was* a force. It acted at a distance. And we said so.

But it was, *because we said so,* that it was true. Einstein's new *description* of gravity doesn't say so. It no longer describes a *single* force. It no longer describes a *single* force acting over a distance.

Specifically, the new description of gravity does not even pertain to a force acting on matter – at all!

Rather, it describes two phenomena: a force due to the presence of matter acting on spacetime; and a measured behaviour of matter responding to spacetime curvatures. We take these *two new* phenomena and call it a force – we call it gravity. Hmmm ...

Note that *how* matter *responds* to curvatures in spacetime is the behaviour we measure – it's what we *call* gravity. The new description says that the force applied by matter to spacetime causes spacetime to bend. But that's not what we measure. We don't provide a measure of how much spacetime is bent and call *that* gravity. Instead, we measure the response of matter to how much spacetime is bent. In other words, our measure of gravity is, at best, indirect. It *was* a direct measurement when we thought it was an invisible attractive force (and some incarnations of our equations still represent it as a direct measurement – because it's useful to do that). But since the description of gravity changed, it now must be an indirect measurement. What we call gravity today is a measurement of how matter *responds* to the produced curvatures in spacetime. Producing curvatures in spacetime is how the force generated by matter manifests – it acts by bending spacetime. But these are two different phenomena – not a single attractive force.

I can't believe that Einstein and others didn't struggle with this *undoing* of the force of gravity, too. But so long as the equations keep on working, and in the absence of anything new to measure, it would have been a brave move for someone to stand up and say that gravity is not a force. I have read that Einstein did say it wasn't, but that got lost in the noise. Today, it is still generally referred to as a force. And because it's a force we can *feel,* that position is easily argued.

And yet, that it remains a force presents its own problems. For example, there is the search for quantum gravity needed to unify quantum mechanics with general relativity. But if gravity did not exist as a *force*, that search would be over. It's difficult to reconcile that, somehow, we knew that there was something wrong with gravity and yet, before first resolving *that*, we attempt to unify it as a force with the other forces we have identified. We can even predict and measure by how much spacetime is bent based on how much mass is present and yet, we didn't think to recast gravity in the context of that measurement. We simply used that measurement to prove that the latest incarnation of the force – gravity as described in general relativity – was correct.

In a way then, what we know today as gravity, involves two measurements.

First, there's a measure of how much spacetime is bent due to the presence of matter. Einstein's field equations quantify that.

Second, there's a measure of the *acceleration* produced in a gravitational Field; the rate at which the speed of two objects travelling toward one another would change, according to a change in their separating distance. This most closely resembles Newton's idea of gravity as an attractive force. With force cast in an equation as mass times acceleration, and having calculated the acceleration and knowing the mass, we can quantify the force and give it a name – gravity.

Given a quantity of mass (our Sun), we can measure how much spacetime is bent due to its mass by measuring how far light veers off course rather than proceeding along a straight-line path. If we were able to take enough measurements, we would be able to see where spacetime was bent and where it wasn't. Of course, we already know it is likely bent everywhere – since 'gravity' acts over an infinite distance – doesn't it?

Newton showed us there was a relationship between matter and matter, which he called gravity – an attractive force.

Einstein showed us, *it was never thus*.

Einstein showed that matter does not attract matter. Instead, he derived the relationship between energy and matter, and then between matter and spacetime. So, in the end, the relationship between matter and matter, *is spacetime*, and specifically, how spacetime is *shaped*. How matter moves depends on the *shape* of spacetime.

So … Ok … show me the force.

It's gone! He killed it!

I hope it is clear now that the idea of gravity as an attractive force between two objects having mass, is obsolete – dead! And not because I say so. But because general relativity itself, in its description of gravity, cannot describe it so. Einstein killed gravity as we knew it. He replaced it with two phenomena. Matter bends spacetime, and matter moves according to how spacetime is shaped.

Our new model will attempt to put all of this into perspective. We will see that Einstein got it right – of course. We will see, not only *why* 'gravity' under general relativity is correct but also why it produced the interpretation that it did. We will tweak that interpretation so that it makes sense. We will see that when we talk about spacetime being *bent* – what that *really* means. We will derive a new interpretation of gravity that is not a force bending spacetime and does not act over a distance. But we will also see how the standard model interpretation came about and why it's still ok to use it – much the same as it's ok to use the Newtonian interpretation for every day earthly events – the 'apple' still *falls* from the tree … right?

Gravity: matter and space

When thinking about what is going on – with gravity – we start by considering the actors in the *play*. We need matter. We need space (we don't need to consider spacetime … yet). That's all we need to bring about the behaviour that we now recognise as gravity. Oh, and we need photons – a light source. We want to recreate the *play* where light (a photon) is travelling through space from a distant star in a straight line toward Earth. But that straight line trajectory is bent when it passes our Sun (matter), due to the Sun's gravity. It is this exact same scenario that was measured to validate the current model of gravity under general relativity in the standard model.

Einstein's map of space and time

What Einstein did with his field equations in general relativity was to create a *map* of spacetime. If you think about an orienteering map, where lines on the map are contours mapping out along the ground where the height above sea level is the same value – well, it's a bit like that. The distance between the lines tells us how far we would need to travel to reach the next line mapping a different height. Where the lines are close together, the slope between one height and the next is steep. Where the lines are further apart, the slope is not as steep. These maps specify for points on the map, their height at that location. So, they show where the point is in three dimensions – the two dimensions of the plane (the surface of the Earth) and the third dimension of height.

If you take two points on an orienteering map, you would be able to compare their height. You would also be able to measure off the map, the separating distance. If the ground between the two points was flat (the same height), then the separating distance would be according to the scale of the map. But if the ground was sloped, the separating distance would be longer than given by the map's scale. That's because the map is drawn in two dimensions on flat paper and a sloped line between two points of different height, slopes upward, *off the map*, forming the long edge of a triangle. That's why we needed the contour lines to show height. Using the difference in height between the two points, we would be able to calculate the separating distance along the sloped path between them. We would also be able to calculate the angle of the slope. We would therefore know at which angle an observer at one point would need to look, to see an observer at the other.

Einstein's field equations map out the terrain of spacetime. In other words, they map out the terrain in four dimensions: the three spatial dimensions, plus the dimension of time. But because time *passes*, this is no ordinary map. If we knew the coordinates of a stationary object in x, y and z, we would know that the object, stationary with respect to the x, y and z axes, would be moving with respect to the time axis.

Now, imagine our orienteering map is of the surface of a lake. The lake is flat. But if we put several fishing boats in the lake, the surface of the lake would not be flat where there were fishing boats. The contour lines would show different heights for the surface of the water where the fishing boats are. We can follow the water line, along the underside of the boat, as a continuation of the lake's surface. We know that the shape of the surface of the lake, that was a flat plane, is changed with the presence of boats.

To use this analogy, what Einstein's field equations would do, is describe the surface of the lake using contour values on the left-hand side of the equation as equal to the shape, weight, speed, and direction of the boats on the right-hand side of the equation. In other words, the shape, weight, speed, and direction of the boats on the right-hand side of the equation determine the shape of the surface of the lake on the left-hand side of the equation at any given point in time. In this analogy, Einstein's approach would be to *equate* the geometry of the lake's surface to the boats shaping that geometry.

In a similar way, Einstein found that the shape of spacetime is changed with the presence of matter. How much matter there is in a region of spacetime determines how much spacetime is bent. And just as a boat changes the shape of a lake's surface, matter changes the shape of spacetime; it creates curvatures along the directions of x, y, z, and time. But an object having matter, even if stationary in the three spatial dimensions of x, y and z, is *moving* through time. So, the shape of spacetime is continually changing, like how the shape of the lake's surface would continually change if the boats were moving. That's how general relativity describes spacetime in the presence of matter. And because matter moves through time, the curvature it causes moves through time as well. And when objects move through space, they are still also moving through time. So, whether an object having mass (or energy), is moving through space or not, because it *always* moves through time, it continually reshapes spacetime. Einstein's field equations equate the geometry of spacetime to the matter shaping that geometry.

But where does gravity factor into this description?

Einstein shows that for each of the dimensions in spacetime, there is an *up* and a *down*. And where spacetime is bent by matter, it is always bent in the down direction. Matter itself would then *fall* through spacetime in the down direction. An object bending spacetime would fall inward – that being the *down* direction. An object encountering spacetime bent by another object would *fall* through spacetime toward where spacetime was bent in the inward direction – the *down* direction. So given the *up* and *down* directions, everything falls - *down*. It's an analogy that we can relate to our everyday experience. Things fall, *down*, not *up*!

Holes in space

The challenge, then, is to conjure up a new way for the path taken by the photon through space, to bend as it passes by a massive object – like our Sun – in a manner like what was observed and measured.

When you think about it, it's not really that difficult to do if we use a little imagination.

The standard model tells us that space is expanding. What if, in the vicinity of matter (having mass), the expansion rate slows down? Better still, what if, in the vicinity of matter, space does not expand at all – but contracts? If we say that when space expands, space it is flowing *into* our universe, then we can say that when it contracts, space is flowing *out* of our universe. We should be able to show therefore, that in a region of space experiencing a contraction, that would present to us as what amounts to the *appearance* of a *bending* of space, were we to regard space as static.

If we deny that space flows (as a fluid might), that it is just there – completely still – then if a contraction *were* taking place, we will measure the effect as a bending of spacetime – and for our purposes here, a bending of space. Let me explain.

When we say that our universe expands, our current understanding is that it does so by stretching. It just gets bigger. A popular analogy is that if space were a raisin cake, then as the cake bakes it gets bigger and the raisins move further apart as it does. We don't say that there is a hole in space through which *additional* space enters our universe causing the expansion by increasing its volume – we are not pumping up a balloon here. But if we allow that space can flow *out* of our universe, that suggests either a reversal of the stretching – a shrinking – or the prospect of a hole in the universe through which space will flow out. All we need to achieve this is to allow that matter (having mass), all particles of matter, are *holes* in space. We will work with the idea that matter does this. Energy will do this, too. Mass (an attribute of matter) and energy are equivalent. But it's easier to visualise what might be going on if we work with matter. We can see it, right?

Where you have a single particle of matter, say an atom (Oops! Can't see that), then that hole in space is quite small. Hence, the amount of space escaping our universe through that atom is less when compared to the amount of space that would escape through a *clump* of matter, say, our Sun. And not all atoms are the same. Some contain more mass than others. So, more space escapes through more massive atoms than through less massive ones. I think you're getting the idea.

So, as a starting point, we want to say that wherever there is matter, all particles of matter are *holes* in space. Because if we say *that*, then we can say that space can leak out of our universe, and the idea of that, as we will see, is very useful.

Space that flows

Let's review. What are we saying here?

We are saying that matter particles, in our universe, are *holes* in space.

But if space is flowing into our universe everywhere causing its expansion and flowing out of our universe through matter, that suggests that space is more than the spatial dimension described in the earlier section, asking 'What is space?' It suggests that space is a fluid. Or at least, that it behaves *as* a fluid.

More importantly, if space were expanding everywhere – through some mechanism – but also contracting locally – because wherever there is matter, there is a *leak* – then that would set up *flows*. Just as we have ocean currents and air currents, the universe, too, would have its own weather system – there would be space currents! Space flowing out of our universe through holes in space due to the presence of matter.

If space were only just expanding everywhere, then space, the ether, call it what you like – would be quite still. We would witness the expansion of space as the distance between galaxies increasing over time as we do. But we would not yet have the bending of space resulting from gravity described in the standard model. But with the addition of space contracting locally, and the resulting space currents, space would not be still. Instead, there would be movement – *everywhere*.

So, with space expanding everywhere and space contracting here and there – *wherever there is matter* – we can see that space, as a medium, would behave as a fluid. And there will be currents!

Bending space without gravity

The chart below depicts the rate at which space is contracting as dashed blue circles radiating outward from the Sun. The further away from the Sun, the slower the contraction. Space contracts more quickly close to the Sun. (Of course, the Sun is not drawn to scale – but larger so you can see it.)

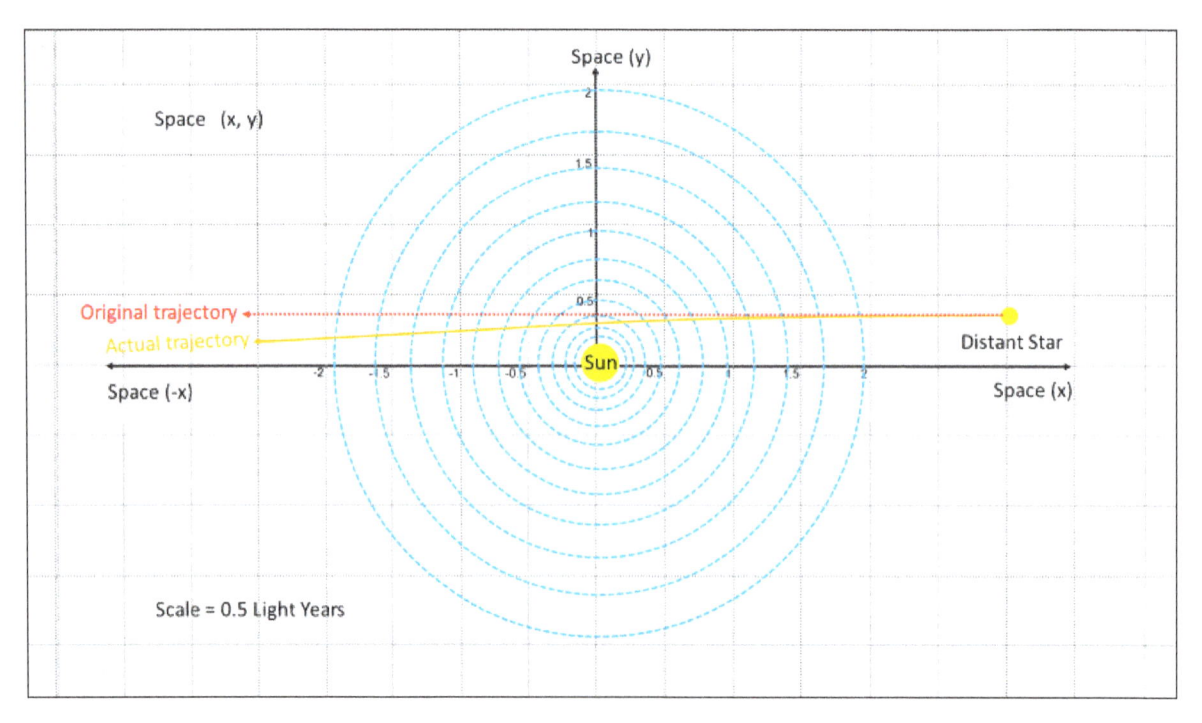

Figure 28.

You can think of the Sun as the drain in a bathtub full of space (space rather than water). As space flows toward the drain, it flows slower further out and fastest when it reaches the drain. Think about a full bathtub at home. When you pull the plug, water flows fastest as it enters the drain and slowest when furthest away from it. If we ran the tap at the same time, filling the bathtub, we would have the volume of water in the tub both expanding because the tap was running, and contracting, because the plug was pulled so the drain was open. If the water from the tap flowed at a faster rate than the water leaving through the drain, our volume of bathtub water would expand over time – even as it is draining where we pulled the plug.

So, what am I saying? What do I propose?

Under the standard model, the phenomenon we call gravity causes space to bend in the vicinity of matter. Light, traveling on a straight-line path through space, finds that when the path goes through where space is bent, its straight-line path is bent as well, because the straight-line path simply follows the space before it – so wherever space is bent, so, too, is the straight-line path. It was this very observation that 'proved' the theory of gravity under general relativity. It proved it because general relativity predicted by *how much* the light would veer off its straight-line trajectory. All we had to do then was measure it when the opportunity came along. We measured it, and the measurements matched the predicted values according to the theory. Astounding stuff!

Having done the experiment, it was concluded that spacetime did indeed *bend* in the vicinity of matter – mass – the Sun. And the explanation for why light veered off course was that its straight-line path through space became bent when that path extended through the region of space bent by the Sun's gravity. Clever stuff! And straightforward, too. All that's required here is to believe that matter bends space – and anyway – we measured it – so it must be true. Right? Well … not necessarily.

What I want to propose is that space *didn't* bend – not at all. Space does not bend. Instead, space flows like a fluid. Space is expanding everywhere. If that's all it was doing, then space would remain still. But if wherever there is matter, where space drains out of our universe, well, that would set up flows and currents.

We would observe a current *indirectly* through the exact same experiment. Light travelling past matter, a lot of it, like our Sun, would be drawn in toward the Sun because the space in the region around the Sun is flowing toward it, into it, and … gone! But because the standard model regards space as still and unmoving, this behaviour is interpreted as spacetime bent in the presence of matter. Under the standard model, we would be able to attribute the bending of spacetime to the presence of matter and we would say that the more matter is present in a region and how dense the mass is, the more spacetime would be bent. Our TRADD model on the other hand, would say that the more matter there is – the more 'holes' there are for space to leak out. When the mass is dense, a great many of these holes are clustered together to make a bigger 'hole'. The bigger the hole, the bigger the leak!

Each particle of matter represents a leak. When particles are packed close together, the size of the leak is the sum of the leaks through the individual particles. The more tightly the particles are packed together, the faster the current.

If water was leaking through a large hole in a bucket, water inside the bucket would need to move toward the hole at some speed. If instead of one large hole we had lots of little holes, the speed with which water would need to move to replace the leaked water at each little hole would be much slower. It is the density of the holes (how close they are together) that determines the speed with which water inside the bucket must move to replace leaked water. Here we are talking about the speed of the current. So, in space, the denser the mass, the faster the current. Of course, we are assuming no change in pressure – no change in the force pushing space out of our universe wherever there's a hole.

Action at a distance

We can now show, how it is, that we would arrive at the idea of a force acting over a distance, if our interpretation of what is going on – what we observe and measure – is a force that bends spacetime.

If we make that force obsolete and say, instead, that what we observe is space flowing out of our universe through holes in space, how would that look? We would say that mass represents those holes so that wherever there is matter, there are holes. A lot of particles of matter in one location would give the appearance of a bigger hole than a single particle. A massive number of particles of matter packed tightly into a region of space might be our Sun. And that would represent a bigger hole than say, the number of particles packed into a region of space that might be Earth. If we revisit the concept of a fluid flowing out through a hole in whatever contained the fluid, we understand that the outflow creates a current. The current is strongest at the hole. But the current still exists a distance from the hole. In fact, the current can exist for a distance away from the hole, approaching infinity.

Let me explain.

Let's say we have a bucket of water that is very tall – say, 10 metres tall – and the bucket is full. The bucket is set on the edge of a cliff with part of the bottom hanging out over the cliff. We then punch a hole in the bottom of the bucket where it overhangs the cliff, and the water begins to leak out. Inside the bucket, a current is set up. Water flowing out through the hole does not leave an empty space inside the bucket. Instead, as the water flows out, water surrounding the hole takes the place of the water that leaked out. And the water that did that does not leave an empty space, either. Water from around that area in the bucket takes the place of that water, too. And so on and so on. When you look at the water level near the top of the bucket, you can see it has gone down. As the water leaks out, the amount of water in the bucket becomes less and so the water level drops. As you can see, water flowing out of the bucket at the bottom, results in water moving inside the bucket. Water at the very top moves down too as the water leaks out. Even if the height of the bucket was a thousand metres tall and the bucket started off full, water leaking out of the bucket would cause the water level to drop. The water at the very top gets to *move* as water leaks out at the bottom. We can almost say that the leak in the bottom of the bucket has set up a *force* able to act at a distance, so that it moves water inside the bucket, all the way to the farthest distance – moving the water at the very top.

The analogy works!

Think of spacetime as a piece of rubber, stretching as the universe expands, and the introduction of matter would produce gravity as we have cast it in the standard model through general relativity. It would be a force acting to bend spacetime and it would act over a distance. The bending of spacetime would result in straight line trajectories of light being bent toward the presence of matter.

Now, instead, think of space as a volume of water, its volume increasing through some mechanism. Then introduce matter as 'holes' to produce flowing space currents. Straight line trajectories of light would be bent toward the 'holes' as space flows out through them. Space flowing out through a 'hole' sets up movement in space that occurs everywhere, not only *at* the 'hole' through which space flows out. But is that really a force acting at a distance? Or is that a local event, influencing things far away from it. When snow on the mountain melts and the water runs downhill into a stream, which runs downhill into a river, which runs downhill into the ocean – is the melting snow a *force* acting on the mouth of the river from a distance?

In the absence of an idea that space is fluid, we are left with the interpretation that space is static, even as it is expanding. It is this perceived static space that is bent by the presence of mass in general relativity. But what happens when we introduce the idea that space is a fluid that can leak out? Then space is no longer static. Then, it's not so much that a static region of space is bent by a force generated by, say, the quantity and distribution of matter in the Sun. Rather, space in the vicinity of the Sun is *flowing* toward it and leaking out. And any straight-line path past the Sun would be *bent* toward the Sun as it sucks in the surrounding space. Yes, the *path* on which the light was travelling did bend. But not because space got *bent* by gravity. Indeed, not because spacetime got bent by gravity. But because the region of space around the Sun was *flowing* toward the Sun. The movement in the ether, the flow of space, space currents, changes the direction a straight-line path would take as it tracks through a region of space where a current is flowing.

There are no holes!

Now, don't get mad; there are no *holes*. That's right, there are no holes in space. Nor are there holes in spacetime. So why, you might wonder, did we go through that detailed explanation about there being holes in space? Well, so that you would understand, if there *were* holes in space, how spacetime could behave like a fluid. And that if spacetime behaved like a fluid, then all the same phenomena that we see and measure when we believe spacetime doesn't behave like a fluid, would still be seen and measured when it does. But the interpretation of what's going on would be different. And, as we will see, the new interpretation of what's going on won't be something magical like matter bending spacetime. We will still be able to make that interpretation, but our understanding of what's going on will be a lot clearer. And it will make sense.

Before we look more closely at *why* and *how,* I'd like to leave you with an analogy to give you a sense of what we mean when we say space is expanding or contracting. Expanding as it moves up the time-axis and contracting due to the 'gravitation' effect – which we'll get to.

Expanding space

Take a rubber band that stretches easily and, using a marker, make three dots on the band, side by side, about a centimetre apart. Now, stretch the band. If you focus on the centre dot, you will see that the dots on either side of it move away from it as the band is stretched. The more you stretch it, the further apart the three dots move. Now, imagine that the dots are galaxies, and the band is space along the x-axis, which stretches as it moves up the time-axis. I think you get the idea. Imagine going up the time-axis at a faster and faster pace. The faster you go up the time-axis, the faster the band stretches and the faster the dots move away from one another. Now the imagine going up the time-axis at a slower and slower pace. The slower the band moves up the time-axis, the slower it stretches. When you stop moving the band up the time-axis the band stops stretching – and time stops still!

Contracting space

You might think now, that for the rubber band to contract, all we need to do is move it down the time-axis. That's not exactly how it works, though. But for the sake of an analogy, let's assume that's what happens, so that we can begin to understand what gravity *is* and how it works. Take a rubber band and stretch it out almost as far as it can stretch without breaking. Using a marker, again make three dots on the band, side by side, about a centimetre apart. Now, as we move the band *down* the time-axis, the rubber band contracts. You can simulate this by allowing the band to contract. As the band contracts the three dots get closer and closer to one another. This is how gravity *works*! If you were standing at one of the outside dots, then as the band contracts, you would be pulled toward the centre dot. The faster the band contracts, the faster you would move toward the centre dot. Later, we will see that the more mass is present (at the centre dot), the faster the band contracts and the faster the three dots on our band simulating the x-axis move toward one another. More mass results in a faster contraction of the band – of the x-axis – of space!

When space contracts

The standard model measures the *force* with which the dots are pulled toward one another as the band contracts – and calls that force gravity. Imagine you could travel along the band from an outside dot away from the centre dot. Then, as the band contracts, you could calculate how fast you would need to travel to maintain your original 1-centimetre distance. In essence, the contracting band you're standing on is pulling you toward the centre dot. You need to run away from the centre dot to hold your ground at 1 centimetre from it. You'd be running away from the centre dot but not making any headway. You'd be running just fast enough to counter the speed with which you're being pulled toward it. When you do that, the outside dot you *were* standing on will still travel toward the centre dot as the band contracts. You maintain your 1-centimetre distance by continually running in the opposite direction.

Leaking space without holes

If we think of the length of the band as the *amount* of space, then while the band is contracting, we see less and less space. Space is leaving our universe due to the contraction.

I hope you can see now, how with our new model, it's possible for space to leak out of our universe through what presents to us as a hole (the centre dot) even as there is no hole. The contraction of space at a point along the x-axis behaves as though there is a hole at the centre dot. And all space left and right of the centre dot is being pulled toward it. As *that* happens, space that was once there – a length of band – disappears – it's gone!

These *contractions* of space occur wherever there is mass. They present to us in our new model as holes in space because they give the appearance of consuming the space around them. If you start with a 100-centimetre band that is stretched and then allow it to contract to 50 centimetres, you get a sense of what I am saying happens in space. After the contraction there is less band. Fifty centimetres of space has gone missing! Where did it go? Of course, we know the band simply contracted. But if we did not know that, then we could interpret what we see differently. We might conclude that 50 centimetres of the band has simply disappeared. Recall our investigations into space – dimensional dynamics? Squash a spatial dimension and its boundary stretches. Reverse squashing that spatial dimension is like allowing the band to contract. You're essentially reversing the direction that you pushed it to make its boundary expand. Do that, and the boundary will contract.

Spacetime is curved ... apparently

You see, what I am about to tell you is that although there are no holes in space or spacetime, there is *something going on* that causes spacetime to behave like a fluid. Yes, that part was true. TRADD contends that spacetime *behaves* like a fluid. It flows. And the reason it does, is because the amount of space in our universe is increasing – the universe is expanding – and because locally (not everywhere), the amount of space in our universe is decreasing. The result is exactly like a bathtub filling with water because the tap is on but with the plug pulled so that water is also draining out. The consequence of that is a *current* due to the water draining. Where the bathtub has one current, in our universe, there are currents everywhere!

You probably thought before that I'd given Einstein a hard time – that I was somewhat dismissive of his work on gravity – on general relativity. Well, as you will see, that was never my intention. In fact, Einstein – in my opinion – got it exactly right! The missing piece, though, is '*how*'. How does matter *bend* spacetime? And given that I am taking the position that matter *does not* bend spacetime, how is it that we see an effect that we *interpret* as matter bending spacetime? And having that interpretation, how does that lead to a different interpretation allowing us to set up a *flow* so that spacetime behaves like a fluid, with currents – like a weather system?

You see, being told that matter bends spacetime and ... *believe it because it's true* ... well, that really doesn't provide people like you and me an understanding of what's really going on. It's like being told that the speed of light is constant. We must believe it's true because that's what our measurements tell us. But that doesn't explain *how* it is that light is always measured to have that same speed – something that will become clear when we look at light later.

OK, so what's *really* going on?

It turns out that matter does play a role in producing a phenomenon. The standard model's interpretation is a bending of spacetime, and when spacetime is bent a lot – it *looks* like a hole in space. You've probably heard it said that if it quacks like a duck and walks like a duck – well, it's a duck! So, what I am about to describe, well, it looks like a hole, stuff falls into it ... like a hole, it looks empty ... err ... like a hole. But it's not a hole. Really, it isn't. But if we *say* it is, then trust me – it will *behave* ... just like a hole. A hole in space. A hole in spacetime, if you like. TRADD says that space *contracts* in the presence of mass. And because of that, we get the *appearance* of space flowing out of our universe just as water flows out of a punctured bucket. But when regarding spacetime as static, as we do under the standard model, we interpret that *appearance* as a curvature in spacetime.

In the standard model, we regard space as static. The bigger the point mass, the stronger the gravitational field. We measure an increasing strength of the gravitational field the closer we get to the point mass. The stronger the gravitational field, the more spacetime is bent – the more space is bent due to the presence of mass. The more space is bent, the stronger the force we call gravity.

In the TRADD model, we regard space as moving – a flowing fluid. The bigger the point mass, the bigger the 'hole' and the stronger the current. We measure an increasing speed of the current the closer we get to the point mass. The faster the current, the faster space moves and the quicker space flows out of our universe via spatial contraction due to the presence of mass. The stronger the current, the stronger the force we call gravity.

When thinking about how gravity is represented in the standard model, consider a massive body moving through space, which it is *bending* as it travels. As the massive body approaches a *location*, space at the location gets more and more bent as the massive body approaches and less and less bent once the massive body has passed and moves further and further away. Is the space along the path travelled by the massive body not moving? Continually? And forever? Does the bending and unbending of space not characterise movement in the fabric of spacetime? In the fabric of space?

If we only measure the strength of the gravitational field with respect to a point mass and the distance from it, we will get the same value each time we take the measurement, regardless of whether the point mass, a massive body, is moving or not. Motion of the massive body with respect to space has no relevance here. We regard space as static. We measure *things* with respect to what we can see *in* space. We measure *things* with respect to other *things*. And when we do that – space remains static. But it's worthy of consideration that as matter moves about in space it is bending and unbending the space it travels through, and that connotes movement, doesn't it? *Movement* – of space – of spacetime, if you insist. Yet we persist with the notion that space, or spacetime, is nothing. We don't measure it. Only the *things* in it. Ugh! We do measure the curvature of spacetime with respect to a point mass. But here again, the focus is on the gravitational field that the curvature generates. And we then attribute that gravitational field to the point mass. To an object. Not to space. In the standard model, everything relates to objects *in* space. Space itself has no currency.

So, the differentiating factor appears to be our definition of space itself; spacetime, if you like. If we say it is *nothing* and void of attributes, then we must regard it as static with respect to mass – with respect to *things*. If we say it is *something*, endowed with attributes, we can then regard it as moving with respect to mass. And therefore, with respect to *things*, we can say it flows like a fluid. When we measure the strength of the force we call gravity, our interpretation of what gravity *is* seems to depend on our definition of what space is, what mass is, and how one influences the other.

So ... let's see what *that* looks like.

Mass in space

The standard model acknowledges the expansion of the universe and has measured its effect on the distances between galaxies as they move further and further apart. Indeed, our measurements of the distances between galaxies over time *lead* to the conclusion that our universe is expanding. Our measurements show that galaxies are not simply moving further apart. Nor are they separating at some constant velocity or at some diminishing velocity. The speed with which they travel away from one another – well, they are not slowing down. They are speeding up! They are accelerating away from one another.

That acceleration decreases with the decline in the mass density of space (as galaxies move further and further apart, the average amount of mass in a volume of space diminishes). This decreasing acceleration sits at odds with the standard model because the model requires a gravitational force to act on the mass of these galaxies to reduce their accelerated recession from one another. Gravity is generated by mass, an attribute of matter, but when we look, we don't see the missing matter required to generate the gravitation we expect is needed to reduce the acceleration of these galaxies.

So, galaxies accelerate. The standard model attempts to define the cause for this acceleration. So far, well, we just don't know. The best we have come up with is a term called 'dark energy'. Dark – because we can't see it. Energy – because all our physics knowledge tells us that for an object to accelerate, there needs to be a force – constantly applied to it – and the application of a force requires energy.

It's not a bad description really. Dark energy. The application of a force we can't identify, see, or measure. But a force none the less. A force responsible for accelerating galaxies away from one another. So, in so much as we measure our universe by measuring *things* in it, when these *things* get further and further away from one another, we conclude that our universe is expanding. We attribute the expansion to a force pushing things outward, away from other *things*. And we call the driver of that force dark energy.

And what of the reducing acceleration of galaxies. Not content to simply infer that dark energy becomes less potent the bigger the universe gets, the standard model instead, came up with yet another term: called dark matter. Yes, as you've no doubt figured out, dark matter is matter that we cannot see as it neither emits nor reflects light. It is needed for more than simply reducing the accelerated recession of galaxies. Galactic systems have been observed to be rotating too fast, given the amount and distribution of the identified mass in the system. What they mean by that, is there is not enough mass in the system to generate sufficient gravitation to prevent the rotating bodies in the galaxies (stars and planets) from flying off into space, given their speed of rotation around the galaxy's centre. Adding matter to the centre of these systems would increase their mass and generate the sufficiently strong gravitational field required to pull on things and hold the system together.

When we look out into the universe and measure *distances* today, we don't factor that the universe is expanding. Surprised? For all practical purposes, when we measure a length between two objects in space, we do that with a rigid ruler and we regard the space between the objects as, well, nothing. It's interesting, isn't it? Space is nothing. But if it is nothing, how can it stretch? OK. Therefore, it does not stretch. It does not move. It is still. Very, very still. But is it? When space encounters mass – then it bends. Come on! Really?

The standard model interpretation stems from the idea that spacetime has no attributes beyond its ability to bend in the presence of mass. There is also the concept of the *vacuum energy*. We won't discuss that here as it's fundamentally a mathematical construct in quantum field theory and in quantum mechanics – and surprise, surprise, these two theories do not agree upon its value. Substantially, quantum field theory requires any value of energy to generate a proportionate gravitational field and the presence of it should be observable – it's not. We don't otherwise measure space. We only measure things *in* space.

Astonishingly, another piece of science we use to confirm that our universe is expanding tells us that space itself expands – it stretches! Yes, that's right. The cosmic microwave background, according to the theory, owes its existence to visible light from the early universe. As the universe expanded over time, space stretched. The visible light in that space – their light waves – stretched, too. Treating light as a wave, as space stretched, a light wave would stretch along with it. The distance between peaks and troughs in the wave increased. The shorter wavelength of visible light radiation became radiation having longer and longer wavelengths over time, until the present day when these waves are so long, we detect them as microwaves.

So, what is our science telling us? How do we interpret what we know? Is space regarded as nothing? Well, most of the time it is. It has no attributes. But then we have the concept of spacetime. And then we have the idea that matter bends spacetime. Additionally, there is the idea that matter can *drag* spacetime behind it as it moves through space – well, as it moves through spacetime. And then there's the idea that space stretches as the universe expands, giving rise to microwaves from visible light, over time. One does wonder what happens to time in these scenarios. Does time stretch, too? And what does that mean? Does it slow down or speed up when it stretches. It slows down, right? So how old is the universe today *really* if since the 'beginning', time has been slowing down due to spacetime stretching?

And although we know our universe is expanding, we say that because we see *things in it* are moving away from other *things*. We have no concept of space having a boundary because we cast the universe as what is called 'closed' rather than open. The surface of a sheet of paper is open. You travel on it to an edge, or you travel on for ever if the sheet has dimensions of infinite length. The surface of a sphere is closed. An ant travelling around the surface of a basketball will never arrive at a boundary. We don't say that space is expanding. We say our universe is expanding but we don't include space or spacetime in that context – well, perhaps, spacetime. I did read that as our universe expands, spacetime stretches – the raisin cake baking in the oven analogy, where the raisins represent galaxies, and the cake represents spacetime. And yet, that stretching of spacetime never really translated into a concept that would infer space is moving. Perhaps it's too difficult to do because we don't have space. We only have spacetime. Because although we measure time, we don't *really* measure space – we only measure distances and when we do that, we only measure distances between *things*. And when we measure a distance – we regard space as being still, unmoving. Very …very … very … still.

In other words, the universe might be expanding, but space 'stands still'.

But does it?

Really?

That's right. It stands still for us when we measure a distance – when we measure a length. Now that makes sense, doesn't it?

Well … not really.

Not according to our new model. In our new model, space doesn't really stand still. But nor, when we measure distance, do we *notice* that it's moving. We don't notice that space *moves*. Well, how could we? Space is *nothing*, right? So how can nothing do anything? How can it move? I know I'm bouncing around here between the standard model where space is *nothing* and our new model where space is *something*. But it's important for us that we first make the shift from *nothing* to *something* if we want to understand an alternate interpretation of what's going on.

One of the things that's going on is that in the presence of mass, space *contracts*. Now that's really going to challenge us after we have just spent time saying that the universe is expanding. Significantly, when the standard model says the universe is expanding, it does not also say that space is expanding – only that the universe is expanding. Well, that's until it wants to talk about the cosmic microwave background. Then, space is expanding – it stretches. And it says it that way because it doesn't measure space – only things *in* it – galaxies, light, and microwaves. So, too, when it says that lengths contract. It is talking about the length of *things* in the universe. It is not talking about a length of space.

But in our new model, when we say the universe is expanding, we say that space is expanding – it is being generated. And when we say that lengths contract where there's gravity, we say that space contracts – it is degenerating. So yes, the interpretation of what's going on in our new model is somewhat different from the interpretation given in the standard model. Where the standard model attributes observations to *things* in space, our new model attributes some of these observations to space itself. And that only works if space is *something* instead of *nothing*. So don't be shy. Try to imagine that space *is* something. We will see that things that didn't make sense before may just start to fall into place once we restore space as an entity with attributes.

Regarding the universe expanding, our TRADD model offers a different interpretation. Galaxies are *pushed* further and further apart due to the expansion of the space between them. Space expands because it is the boundary of a hypersphere that is undergoing a deformation, expanding its boundary. Analogous to the standard model, dark energy would be the driver of the force responsible for the deformation.

Introducing a point mass

Let's look at our model so far. We are saying that the x-axis is moving upward through the time-axis and that, *therefore*, time passes. We are also saying that as the x-axis moves upward, it stretches, and because of that, our universe (our space in x, y, and z) expands. But what would happen if we added some matter? An attribute of matter is mass. I will use the term 'point mass' to describe a point in space, along the x-axis, where there is mass. So, for our model, what would happen when we place a point mass on the x-axis?

In 'chart 1', we place a point mass on the x-axis at x = (-1). We are using a scale able to show us what is going on – which is not necessarily how things would scale. We have already discussed and shown how the x-axis stretches as it moves upward through the time-axis. So, in this chart, the x-axis is at some point along the time-axis, and we have rescaled both axes so that the x-axis cuts the time-axis at zero. We know from the standard model that two things should happen due to the presence of the point mass. Lengths contract and time slows down. These changes in the presence of mass have been measured and validated under the standard model. Hence, our new model ought to account for them. Remember, we do not seek to debunk the standard model – but instead, show what is going on in the universe from a different perspective and in a way that makes sense.

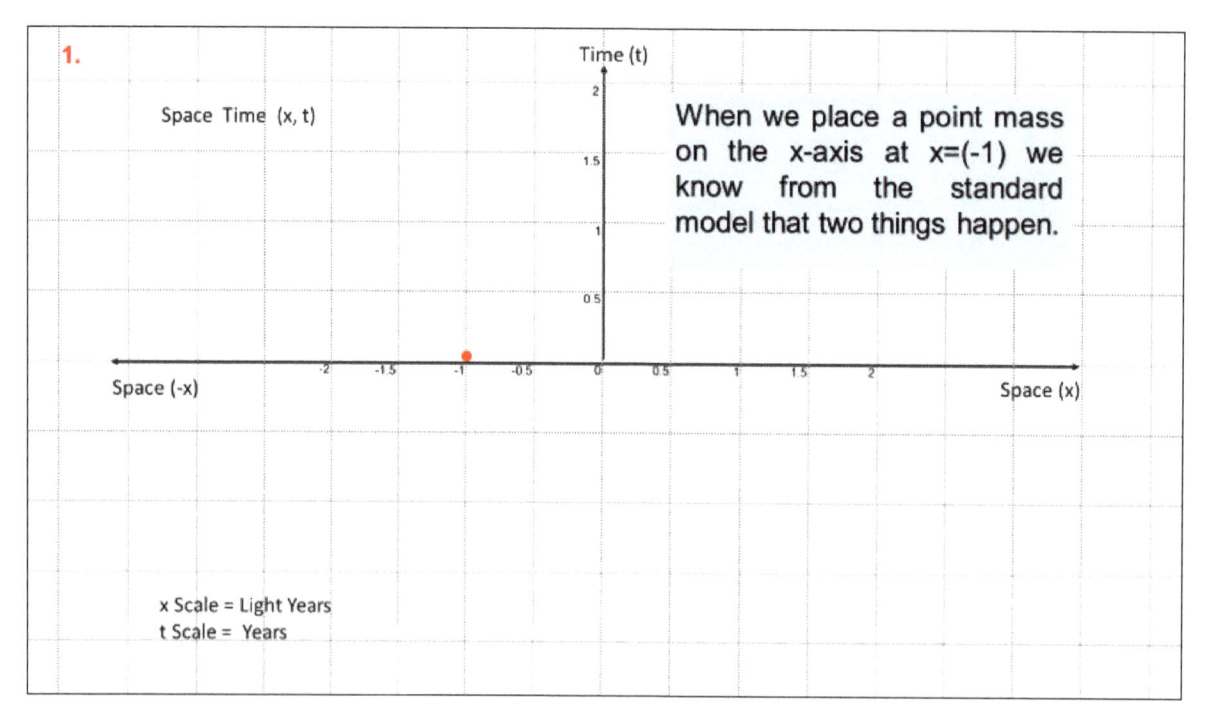

1.

Figure 29.

Space Time (x, t) graph with Time (t) on vertical axis and Space (-x) / Space (x) on horizontal axis. A point mass is shown at x=(-1).

"When we place a point mass on the x-axis at x=(-1) we know from the standard model that two things happen."

x Scale = Light Years
t Scale = Years

Figure 29.

Lengths contract

Our model tells us that as the x-axis moves upward through the time-axis, it stretches – it expands. So, in that respect, length, along the x-axis, is increasing – continually.

So how do we show length contracting?

If the x-axis expands at some rate in the absence of a point mass, then the introduction of a point mass should slow the rate at which the x-axis expands. Then using a rigid ruler at some instant in time, we could measure distances along the x-axis where the expansion rate has slowed and compare them with distances along the x-axis where the expansion rate did not slow – where there is no point mass. We would find distances become shorter as we approach the point mass and become longer moving away from the point mass. In effect, lengths in the vicinity of the point mass contract. And the closer to the point mass, the larger the contraction.

2.

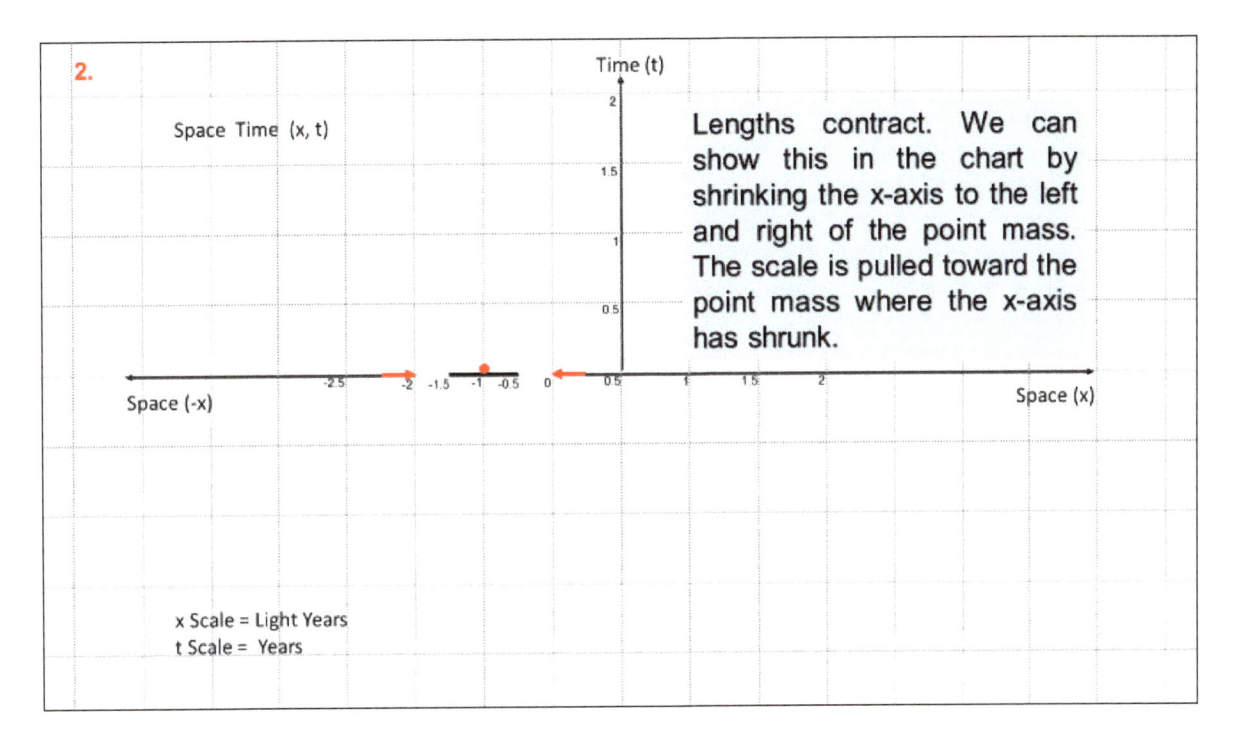

Figure description: Space Time (x, t) chart with Time (t) vertical axis and Space (x) horizontal axis. x Scale = Light Years, t Scale = Years.

Lengths contract. We can show this in the chart by shrinking the x-axis to the left and right of the point mass. The scale is pulled toward the point mass where the x-axis has shrunk.

Figure 30.

Time slows

When we introduced 'time' for this model, we said that: 'Time is the number of years travelled by the x-axis in the time-direction'. We arrived at this conclusion by first considering the fourth dimension as a direction. We called that the w-direction. We then showed that by renaming it the time-direction and re-scaling it from metres to seconds, we could represent 'time' as the movement of the x-axis upward through the time-axis.

But we didn't really go into any detail about how fast (or slow) the x-axis would move upward through the time-axis. One would be excused for thinking that the x-axis moved up through the time-axis at a rate of 1 year, every year. To think otherwise would imply a rate requiring a second clock – another clock. We would have to say something like – the x-axis moves up through the time-axis at a rate of 1 year every 6 months.

Wow!

That would be confusing without some explanation. But I must tell you, that's exactly what happens. Not the bit about the *6 months*. But that the x-axis does not always travel upward through the time-axis at the same speed. As it travels, it does pass the 6-month mark, and the year mark, and so on. But how fast it moves varies. And mass is one of the reasons for that variation in speed.

I will explain more about 'time' later, including little details like – *the x-axis does not always travel at the same speed*. It would have been confusing to explore the *speed* of the x-axis when introducing 'time'. That's because 'time' is closely linked to several phenomena that each require more explanation. As those phenomena are discussed and their links to 'time' explained, a fuller explanation of 'time' will emerge. For now, we can augment our understanding of 'time' by noting that the speed at which we move through it – the speed of the x-axis – can vary.

To show that time slows down in the presence of mass, we need to slow the progress of the x-axis upward through the time-axis. So, where there is a point mass, the x-axis moves a little slower. The more mass at the point, the slower the x-axis moves as compared with its movement elsewhere. So, where time slows in a region along the x-axis, that part of the x-axis *lags* the rest of the x-axis upward through the time-axis.

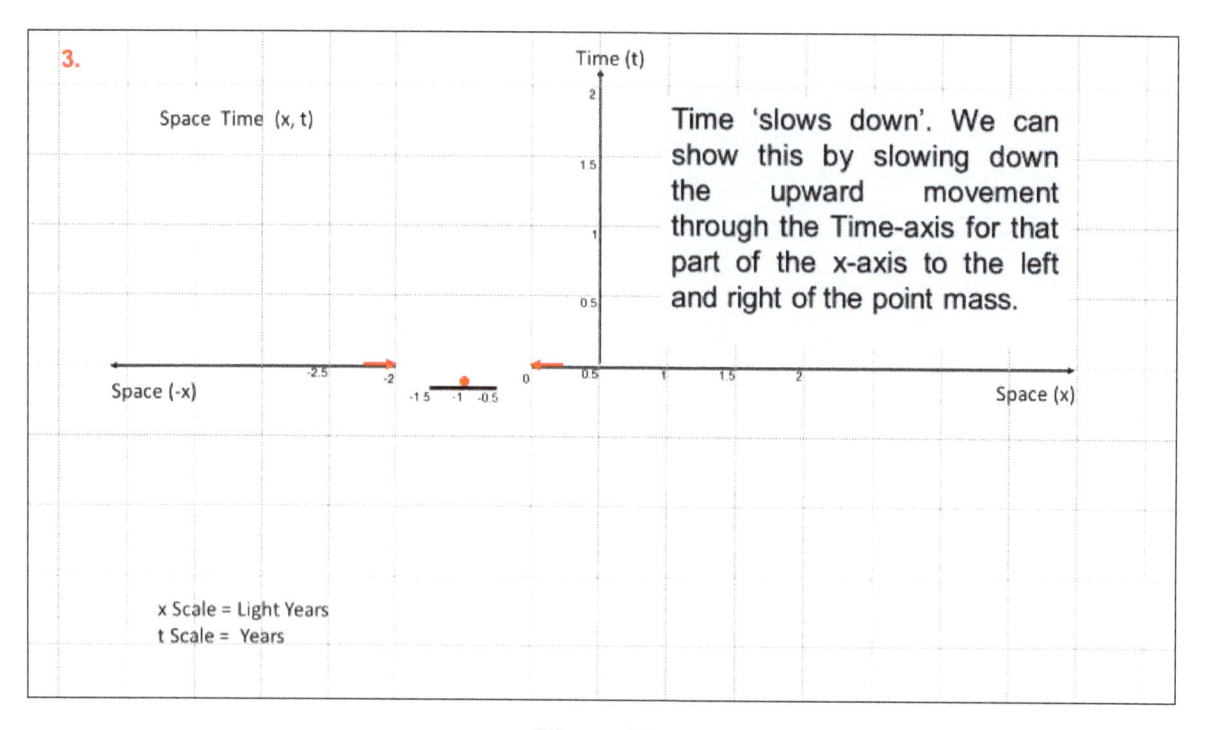

3.

Space Time (x, t)

Time 'slows down'. We can show this by slowing down the upward movement through the Time-axis for that part of the x-axis to the left and right of the point mass.

Space (-x)   Space (x)

x Scale = Light Years
t Scale = Years

Figure 31.

Proximity smoothing

Wherever there is a point mass, the progression of the x-axis upward through the time-axis is slowed. The closer we are to the point mass, the more pronounced the *drag* on movement of the x-axis. To show this, we need a curved line cradling the point mass where there is the most drag. Then we need to show reducing drag as we move away from the point mass to the left and right along the x-axis.

As the drag lessens, the expansion rate of the x-axis increases.

We expect that the closer we are to the point mass, the more pronounced these effects are. We show a reduction in these effects moving away from the point mass.

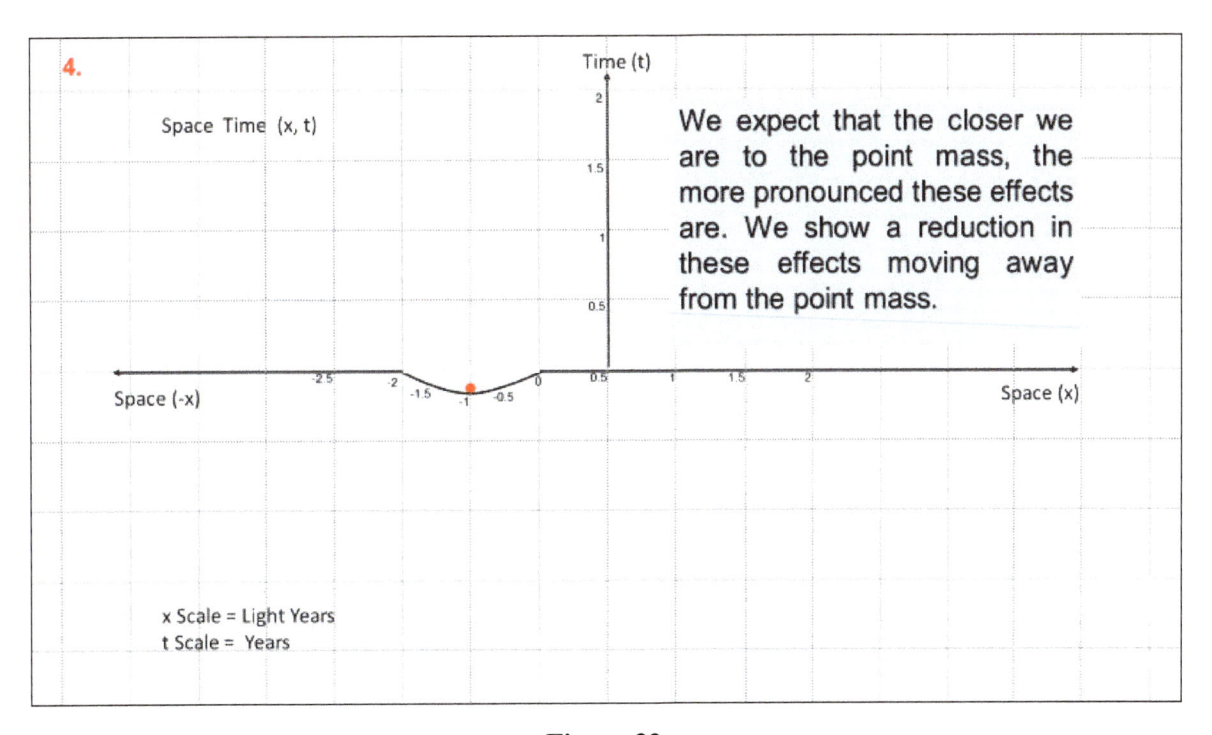

Figure 32.

First time, then length

Let's look at what we have done so far.

Chart 1. We introduced a point mass at -1 on the x-axis.

Chart 2. We show the effect of length contraction on the x-axis.

Chart 3. We show the effect of time slowing on the x-axis.

Chart 4. We *smooth* the effect of length contraction and time slowing on the x-axis.

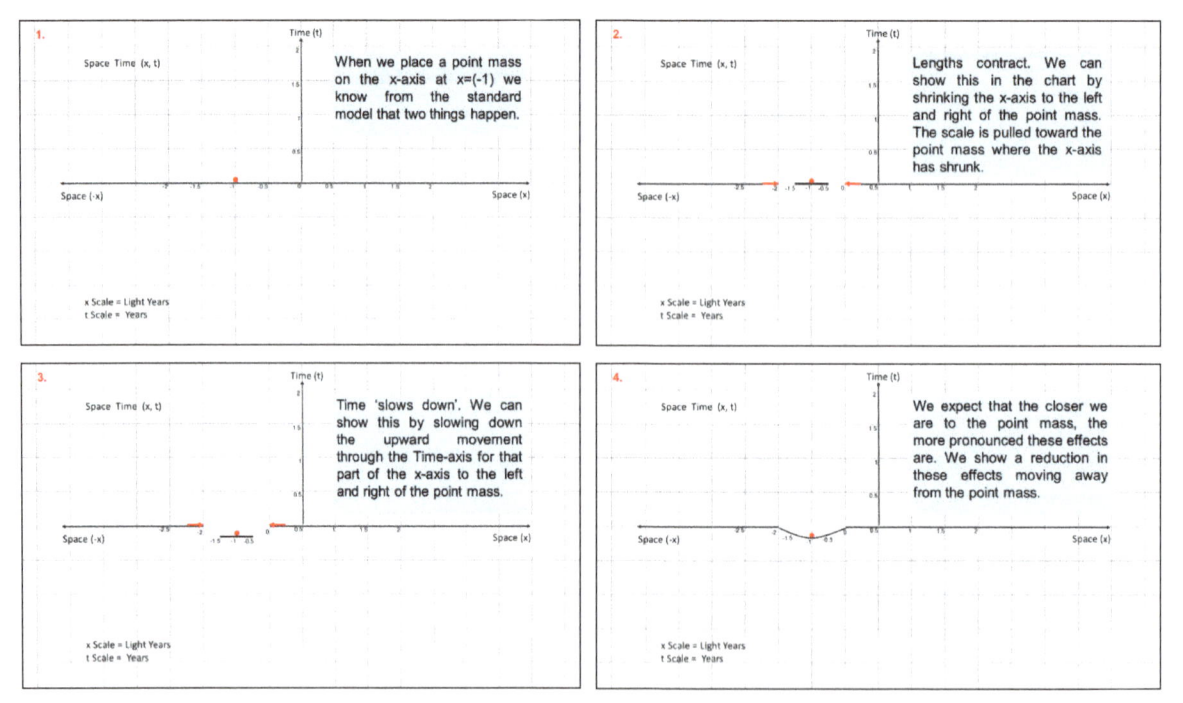

Figure 33.

But does this tell us what is happening?

It describes how the model would accommodate *length contraction* and *time slowing down*. But it doesn't really tell us what is happening.

Does length contracting due to a point mass occur independently of time slowing down?

Does the presence of a point mass cause both length contraction and the slowing of time?

Or does time slowing down influence length contraction or is it the other way around; does length contraction cause time to slow down?

Recall that we have been exploring length contraction and time slowing down because the standard model tells us that these effects occur in the presence of mass. Our aim was to see how we could show these effects in our new model. The best way to do that was to deal with length contraction first so that we could depict the contracted length being pushed down away from the rest of the x-axis to show the effect of time slowing down. We did that. But now we also want to understand what is causing these effects.

Earlier, we said that the universe was expanding – at a rate. Without immediately jumping to what that rate might be, that there *is* a rate, tells us that as our three-dimensional sphere moves through the time-axis, (in the w-direction), two things (so far) are changing. We viewed what is happening using the x-axis from our sphere and the time-axis to get a clearer picture of events.

1. Progress of the x-axis upward through the time-axis
2. Expansion of the x-axis

Development of the model tells us there is a *force* pushing on the boundary of a hypersphere – squashing it. The boundary of a hypersphere is a volume. The x-axis is a component of that boundary and so when we take an x-axis and time-axis view, we are saying that the *force* is pushing the x-axis upwards through the time-axis. We can also conclude that the introduction of a point mass slows that pushing, and hence, that the point mass influences the strength of the pushing force or the resistive force. We can cast that influence as a weakening of the pushing force or a strengthening of the resistive force – or a little of both.

We know the expansion of the x-axis is a *consequence* of the axis moving upward through the time-axis. On introducing a point mass, we dealt with length contraction first and time slowing down second, because it was just easier to show what was going on that way. So, let's see what happens when we deal with time slowing down first and length contraction second. We want to reverse the order to see if that gives us a clearer understanding. So, let's do some diagrams again, but this time in a different order.

Introducing a point mass again

In chart 1, we place a point mass on the x-axis at x = (-1). We are using a scale able to show us what is going on – which is not necessarily how things *would* scale. This time around, we have included some of the forces identified earlier. We have only just placed the point mass. We expect that the x-axis at the point mass would move up the time-axis at a slower rate than further away from the point mass. Therefore, the point mass on the x-axis would move up the time-axis at that same slower rate.

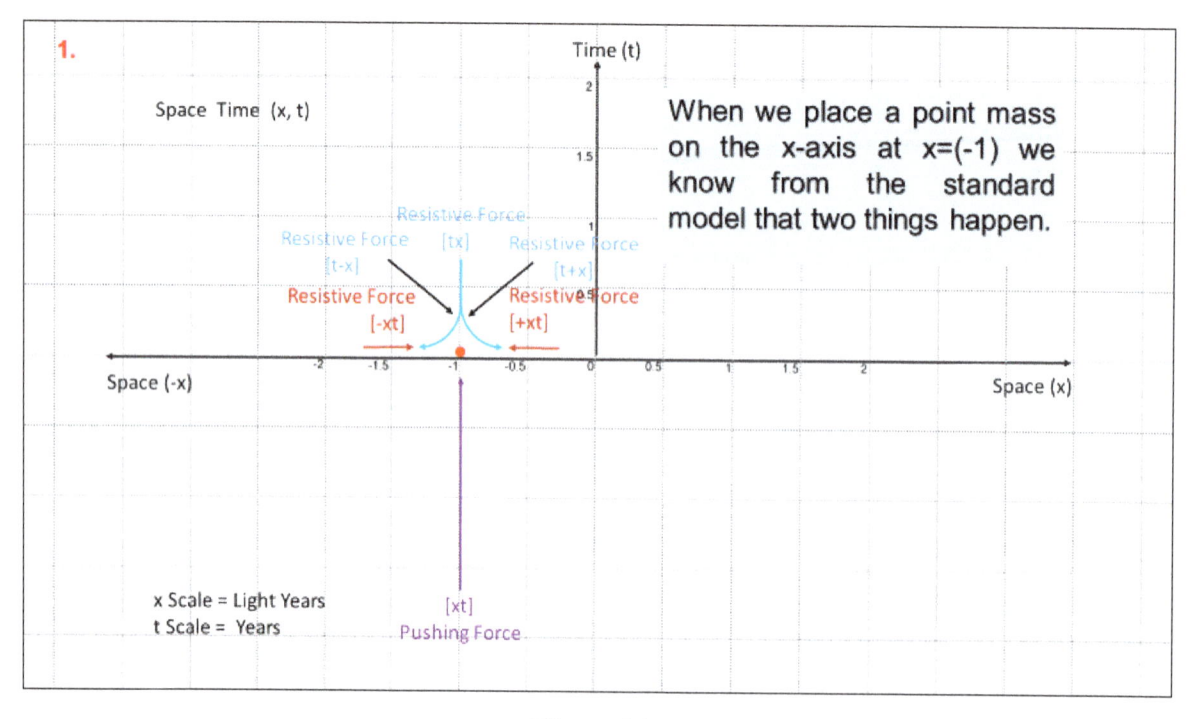

Figure 34.

Mass slows time

In chart 2, we show that the point mass is lower down the time-axis than the x-axis. It is lagging the rest of the x-axis, which continues to move up the time-axis. We have not yet adjusted for any expansion or contraction along the x-axis – we have just left that section of the x-axis blank for now. We did rescale the time-axis so that the x-axis cuts it at t = 0. It is important to note that at the point mass, the x-axis is still moving upward along the time-axis but because it moves slower at the point mass than remote to the point mass, it is lower against the time-axis at the point mass.

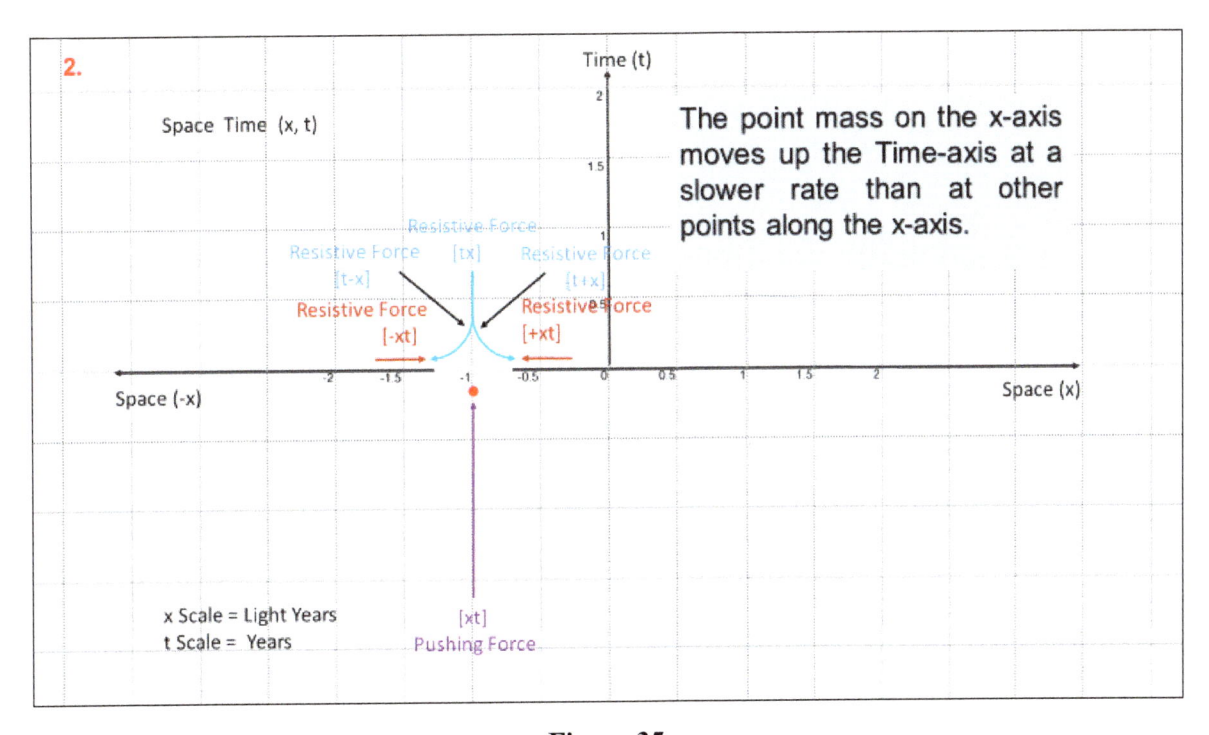

Figure 35.

More mass slows time more

In chart 3, we show that the point mass is moving up the time-axis at an even slower rate. That's because we increased the size of the point mass. The standard model tells us that the more mass is concentrated in a region of space, the more time will slow down. Here, we show the point mass lagging even further behind the rest of the x-axis than when the point mass was smaller. We have not yet adjusted for any expansion or contraction along the x-axis – we have just left that section of the x-axis blank for now. We did rescale the time-axis so that the x-axis cuts it at t = 0. But it's important to remember that the x-axis at the point mass and remote to the point mass is moving up the time-axis. It just moves upward more slowly at the point mass than remote to it.

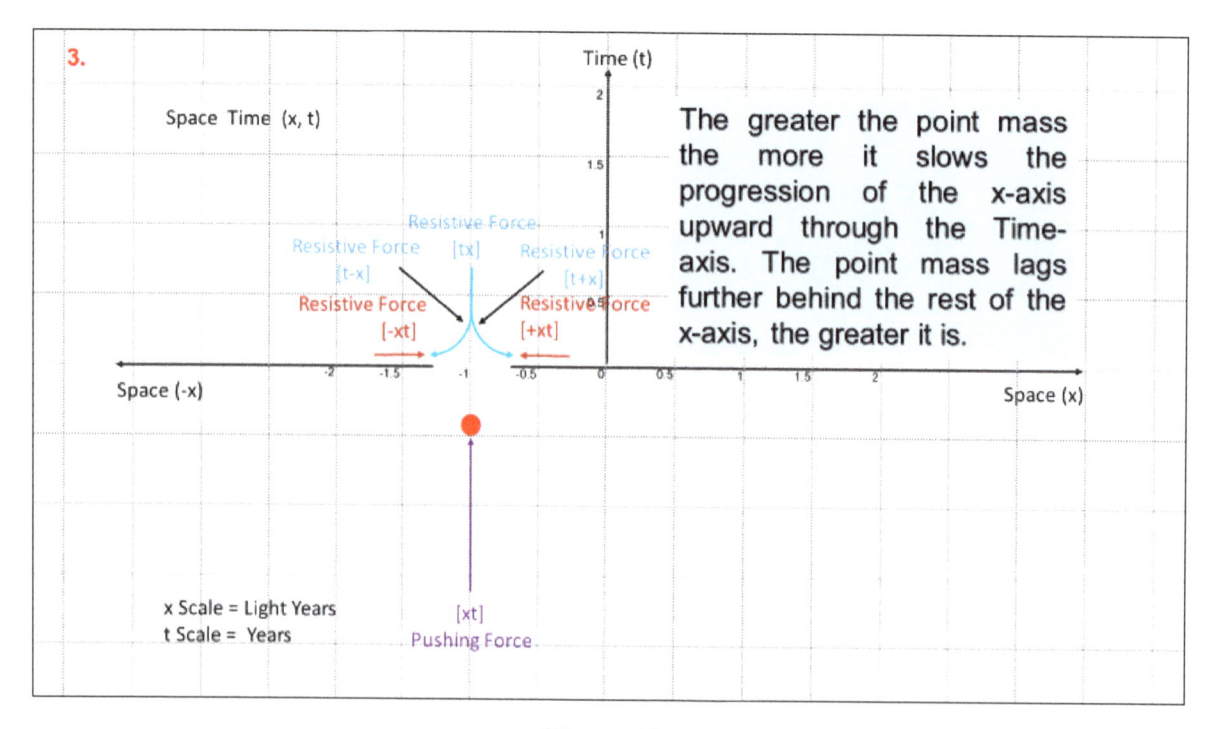

Figure 36.

In chart 4, we show that the speed at which the x-axis moves upward through the time-axis is influenced by the size of the point mass, with no point mass giving the greatest speed. The larger the point mass the slower the speed. We show using the red markers how much time has slowed down for a small point mass as compared with a larger point mass. We then show using green arrows, how much the x-axis would expand depending on the speed of the x-axis upward through the time-axis. Because the expansion of the x-axis is related to the movement of the x-axis up the time-axis, the faster the x-axis moves, the faster the expansion. When considering a point mass, the larger the point mass, the slower the speed of the x-axis up the time-axis and hence the slower the expansion.

So, we conclude that a point mass influences the relative strengths of the xt pushing force and the tx resistive force. That in turn determines how fast the x-axis moves upward through the time-axis, which in turn determines how much the x-axis expands. We see that local to the point mass the expansion of the x-axis is slowed as compared to its expansion remote to the point mass. What is different about the expansion at the point mass and a point on the x-axis remote to the point mass is how fast the x-axis moves up the time-axis: how long it takes for time to pass.

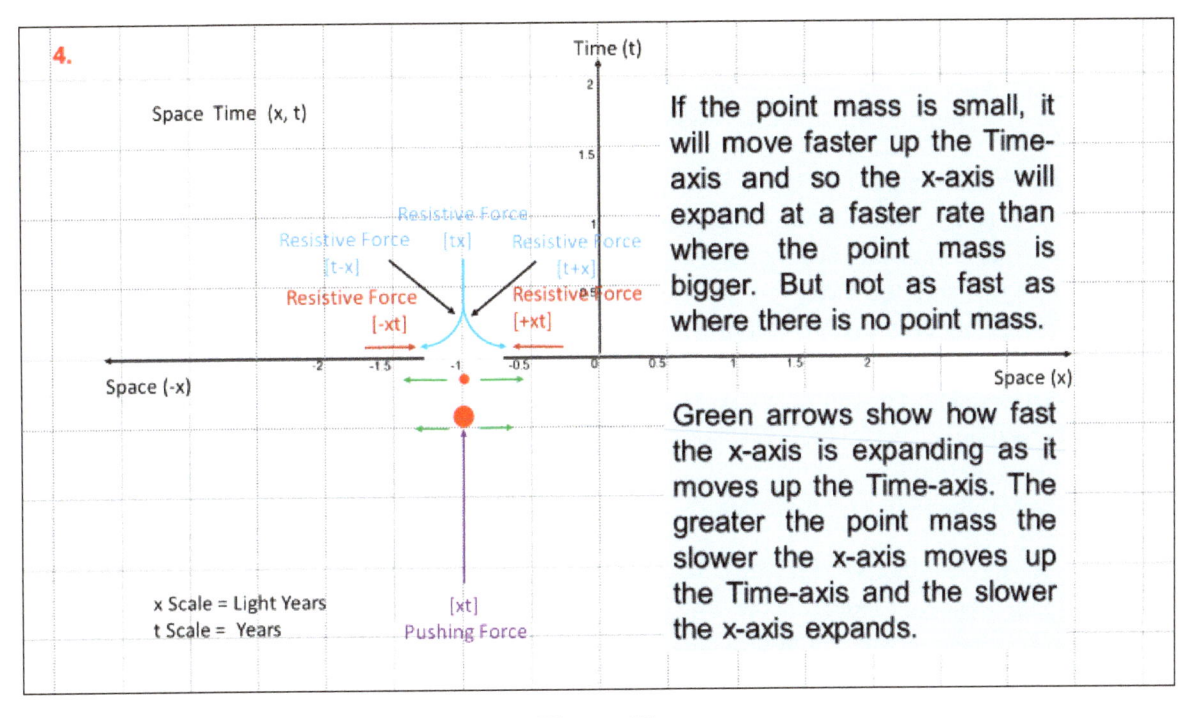

Figure 37.

Having dealt with time slowing down by showing it is the slowing down of the speed with which the x-axis moves upward through the time-axis, we want to deal with length contraction next.

Contracting space contracts lengths

In chart 5, we show several things. First, we say that the expansion rate of the x-axis is n-meters per year. That's a way of saying that if we don't know by how many meters per year space will expand, we can just use a symbol to denote some unknown number of meters. So, we use n-meters.

We know that the expansion rate depends on how fast the x-axis moves up the time-axis.

But what we find is that where progression up the time-axis is slower due to a point mass, the expansion rate remains at n-meters per year.

In the model, a year is defined as the x-axis moving upward through the time-axis for 1 year. So, if it moves upward for 1 year *quickly* because there is no point mass slowing it down, or if it moves upward for 1 year *slowly* because there *is* a point mass slowing it down, the result is the same. It moves up the time-axis at a slower speed, for a month, for 6 months, and eventually for 1 year, after which it will continue to move upward. And having moved up for 1 year, the x-axis will have expanded by n-meters. Get it? N-metres per year.

This extraordinary result underpins much of what the model reveals about space and time. The expansion rate of the universe, in any region of space, whether remote to a point mass or at the location of a point mass, and regardless of the size of the point mass, is the same: n-metres per year.

You may want to think on this for a few moments to fully grasp its significance.

Also, to help you visualise the expansion, recall that the x-axis is the diameter of a sphere. So, when we say it expands by n-meters, we should apply that, strictly speaking, to the length of the diameter. But it's a cleaner solution to apply it to the radius, as an expanding sphere does so by extending its radius. Hence, if the radius extends n-meters per year, the diameter will extend 2 n-meters per year. It doesn't really matter which measurement we use at this point because we don't know the value of 'n'. But I did want to give you some sense of what we mean, in the context of the model, when we say that the x-axis expands (or contracts). And … yes, there's more. You'll recall that our universe in the context of our model is not merely a sphere – but a sphere in the boundary of a hypersphere. And as we kind of demonstrated earlier – that's an object that we have difficulty visualising. So, to sum up – the x-axis here is the radius (and diameter) of a sphere component of the boundary of a hypersphere.

The TRADD model suggests it is *this* relationship between space and time that gives rise to the assertion in the standard model that the laws of physics are everywhere the same. Our new model tells us that the mechanics of space and time, the way space behaves with respect to time, never changes. It's a constant ratio. It's n-metres per year, where the value of 'n' is a constant. Space *always* expands at n-metres per year. And as we will discover in the section on light, the TRADD model compels us to consider that the measured speed of light in a vacuum is also related to n-metres per year. But we're looking at gravity here. And, as we will see, when mass is involved, we need to take a few more things into account.

Let's continue looking at chart 5.

The expansion rate to the right of the time-axis in the chart below is n-meters per year. A distance spanned by the green arrow has a length of 1 light-year on the right, but that same *distance* has a length of 2 light-years on the left (spanned by the light blue arrow). That is because the point mass on the left slowed the speed of the x-axis upward through the time-axis and so that part of the x-axis did not expand as quickly as it did on the right side of the chart. The scale on the x-axis appears compressed near the point mass as compared to the x-axis remote to the point mass. Yet, the expansion rate on the left side of the chart is the same as on the right side of the chart. It is n-meters per year. But it takes a little longer for a year to elapse on the left side of the chart because the point mass has slowed the speed of the x-axis upward through the time-axis.

From the perspective of an observer on the right side of the chart, the length of the green arrow on the left side of the chart has contracted – become shorter. And that's what the standard model is telling us – that in the vicinity of mass, lengths contract – and that's exactly what our model shows. But in our model, we can also see what is going on. We can see *why*, from the perspective of any region in space, an observer looking at a different region of space where there is more mass, will measure that the lengths in that region have contracted. Importantly, the observer will be able to attribute the contraction of lengths to the amount of mass present in the region. Also, the observer will note that time has slowed in the region and can again attribute that to the amount of mass present.

Of course, the observer on the right side of the chart does *not really* see the green arrow on the left side of the chart. It is the light blue arrow that is seen. And having measured its length from afar as being 2 light-years, half its length would derive the green arrow spanning 1 light-year.

5.

Space Time (x, t)

2 Light Years

1 Light Year

1 Light Year

Space (-x)   Space (x)

Let's say that the x-axis expands at a rate of n-meters per year in a region remote from any point mass.

Where there is a point mass the expansion rate is still n-meters per year. But each year will take *longer* to elapse.

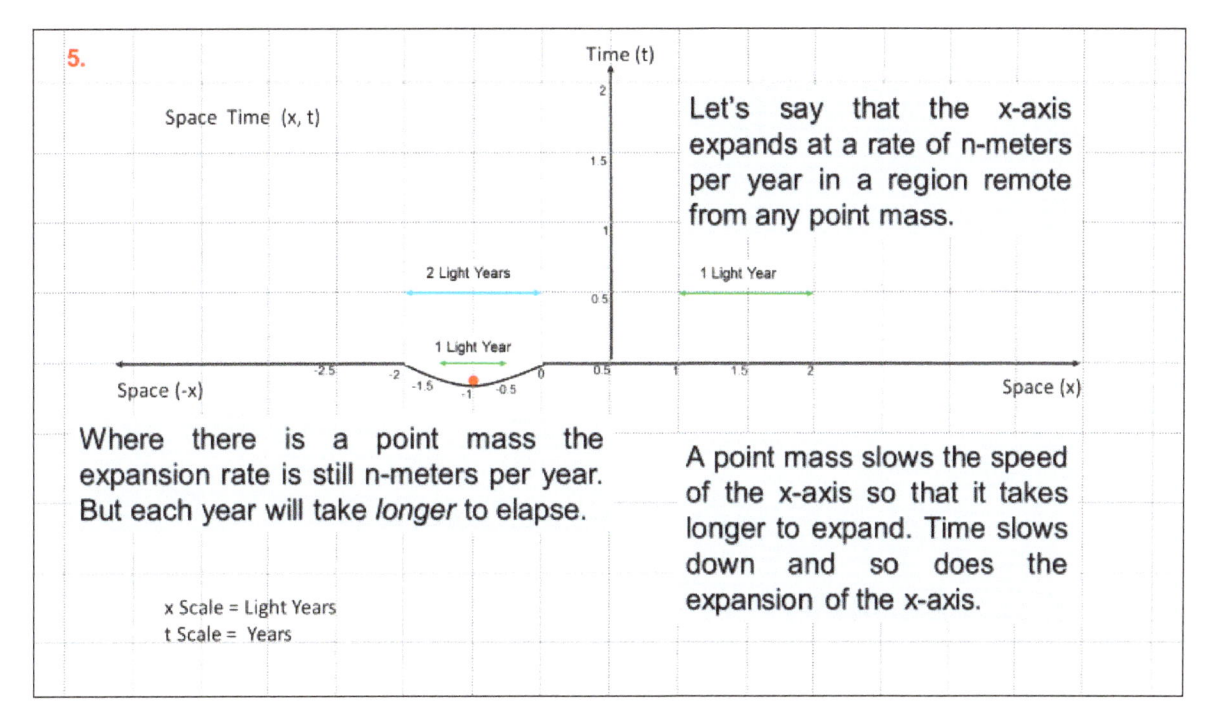

A point mass slows the speed of the x-axis so that it takes longer to expand. Time slows down and so does the expansion of the x-axis.

x Scale = Light Years
t Scale = Years

Figure 38.

So, the presence of mass when observed from a region in space having less mass, will produce measurements showing time has slowed down and lengths have contracted, according to and proportionate to, the amount of mass present in the region being observed.

Mass changes time changes space changes length

Let's look *again* at what we have done - so far.

Chart 1. We introduced a point mass at -1 on the x-axis.

Chart 2. We show that mass slows time.

Chart 3. We show that more mass slows time more.

Chart 4. We show that slowing time contracts space.

Chart 5. We show that contracting space contracts lengths.

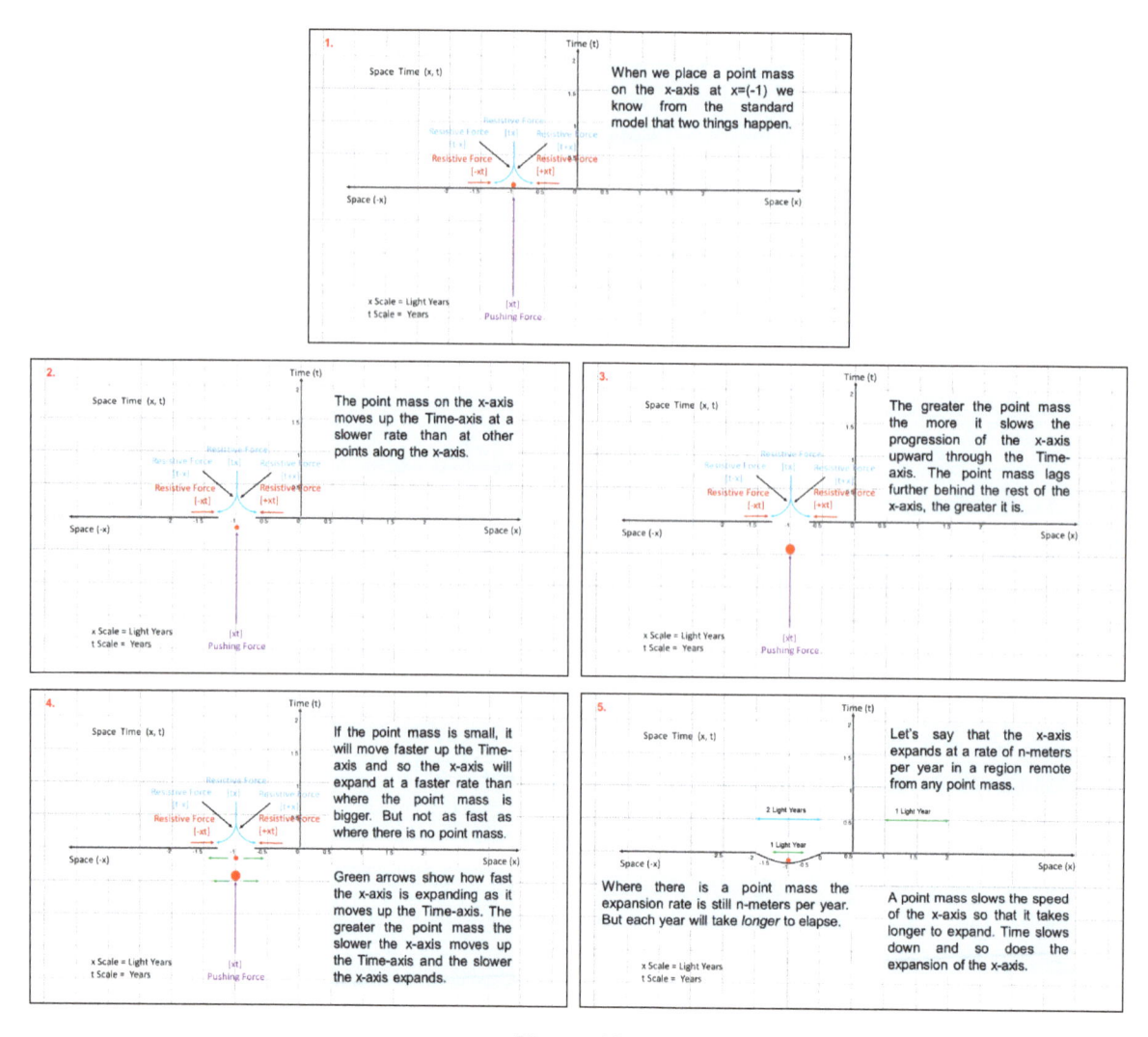

Figure 39.

OK, so where is gravity?

But hang on!

Aren't we forgetting something?

The TRADD model says that the x-axis expands because it is moving up the time-axis. We have shown a change in the speed that the x-axis moves *up* the time-axis due to the presence of a point mass. A slower speed. But we described gravity as a contraction of the x-axis. Not a slower expansion. Sure, an observer remote to the point mass will view the space at the point mass as contracted. Lengths at the point mass are shorter. But we don't see gravity at work here – not at all. And that's because once again, we have fallen into the trap of regarding space as static! We are measuring lengths and relative lengths as though there is no movement in the x-axis. Sure, we state that the x-axis is expanding as it moves up the time-axis. So, there's movement, right? Yet we measure the relative lengths of space at the point mass and remote to it as static. No, for gravity to be at work, for us to feel the force of it, space needs to move. It needs to contract. When it contracts, all points along the x-axis will be pulled toward one another, just as points on a rubber band would be if it were allowed to contract after having been stretched.

That's gravity!

So where is it?

The standard model, which we want to honour, tells us that lengths contract in the presence of mass. But under the TRADD model, the x-axis always *expands* as it moves up the time-axis. Don't we need for the x-axis to move *down* the time-axis for it to *contract*? Don't we need to reverse how space expands to show how it contracts?

No!

The solution we seek is a little subtler than that.

We were forgetting something.

Force.

Mass mediates force

In chart 1, we place a point mass on the x-axis at $x = (-1)$. We are using a scale that can show us what is going on – which is not necessarily how things *would* scale. Again, we have included some of the forces identified earlier. We have only just placed the point mass. We expect that the x-axis at the point mass would move up the time-axis at a slower rate than further away from the point mass. Therefore, the point mass on the x-axis would move up the time-axis at that same slower rate.

But now we are also interested in what happens with the forces at the point on the x-axis where the point mass is placed.

Let's review what these forces are.

Recall that in the section on dimensional dynamics, we saw what happens when a dimension is squashed (deformed), and its defining attribute is conserved. Hyperspace is the defining attribute of a hypersphere, and it is bounded by a volume, the next lower dimension.

The force, which we named [xt], is pushing against the boundary represented by the x-axis, deforming the hypersphere. There is resistance to [xt] from the force we named [tx]. The resistance is a manifestation of the hypersphere's effort to conserve its hyperspace. It's analogous to a constant pressure applied by the hypersphere to its boundary. We can say that all its hyperspace exists at some measure of pressure and that pressure therefore presents at the boundary. But because the deformation does take place, we know that the force [tx] cannot match the force [xt]. It can't resist it. Instead, [tx] divides into [t-x] and [t+x], pushing perpendicular to [xt] along the x-axis in both directions, stretching it. It's not unlike the surface of an inflating balloon, stretching because it cannot match the air pressure pushing against it from the inside.

But the x-axis responds to resist the stretching forces of [t-x] and [t+x], pushing back with forces [-xt] and [+xt]. The resistance results in the x-axis stretching at a *rate*. If you pull both ends of a rubber band, it stretches. But there is resistance to the stretching.

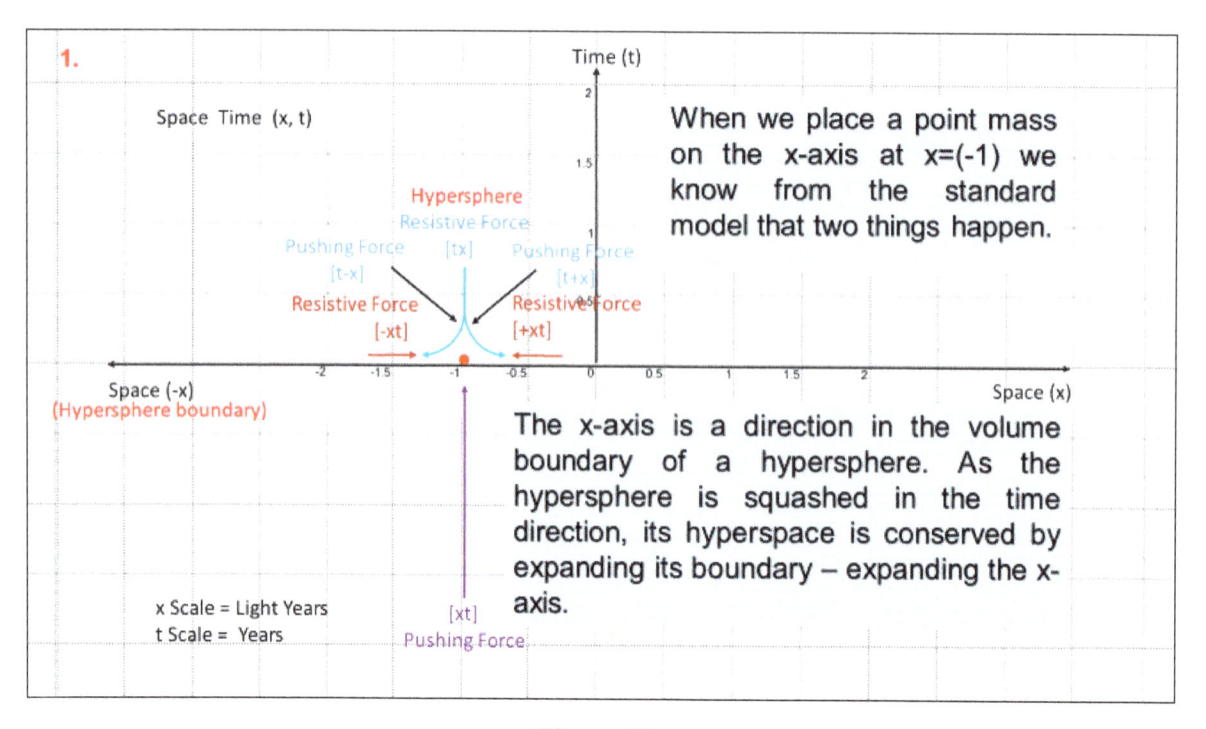

Figure 40.

Mediated force slows time

In chart 2, we show what appears to be the point mass moving down the time-axis. But it is important to understand that it isn't *really* moving *down* the time-axis at all. It's moving up the time-axis, but at a slower speed than the x-axis moves up the time-axis remote to the point mass. At the point mass, the x-axis is lagging the rest of the x-axis. The entire length of the x-axis is moving up the time-axis. We have not yet adjusted for any expansion or contraction along the x-axis – we have just left that section of the x-axis blank for now. We did rescale the time-axis so that the x-axis remote to the point mass cuts it at t = 0. And that's why it looks as though the point mass moved down the time-axis – because we rescaled the time-axis.

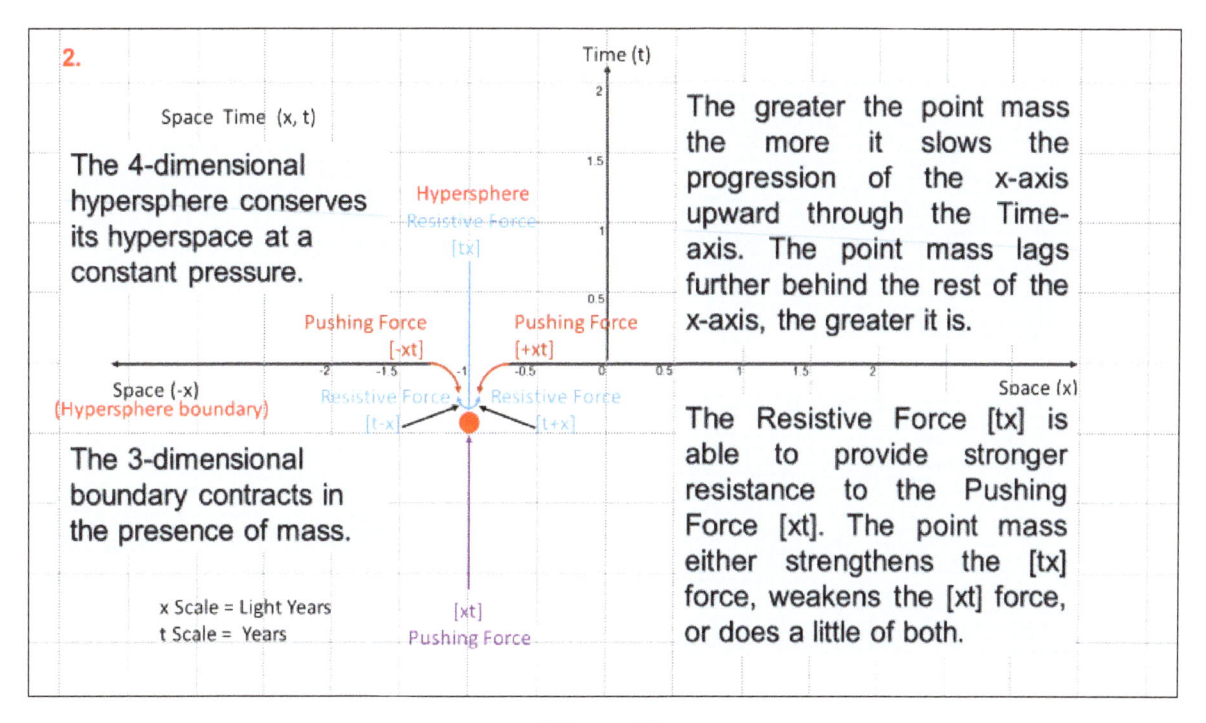

Figure 41.

But recall that before placing the point mass, it was the resistive forces [t-x] and [t+x] that were responsible for stretching the x-axis at every point along it. And resisting these were [-xt] and [+xt]. The resistive forces [t-x] and [t+x] were residual to the Resistive Force [tx]. Where [tx] was able to resist the Pushing Force [xt] to some extent, it was not strong enough to resist it completely. That's why the x-axis gets pushed upward along the time-axis. But where [tx] was unable to completely resist [xt] head on and directly, it divided, redirecting some of its force into [t-x] and [t+x], in opposite directions along the x-axis, stretching it.

Now, though, with the point mass present, [tx] can offer increased resistance to [xt], slowing the progression of the x-axis up the time-axis.

Mediated force contracts space

But in offering increased resistance, less force remains for use by the residual forces [t-x] and [t+x]. They lose power. But the strength of the resistive forces along the x-axis, [-xt] and [+xt] remain the same. Whereas before, they were slowing the expansion induced by [t-x] and [t+x] that were pushing points along the x-axis apart, stretching it, they can now overwhelm [t-x] and [t+x], pushing points along the x-axis together again, contracting it.

Contracting space is gravity

And there it is.

Gravity.

Contracting space!

So here we have a special case for the expanding boundary of hyperspace in the presence of and due to a point mass. In the presence of mass, the x-axis contracts, even as it is moving forward through time – up the time-axis – albeit at a slower speed than the speed of the x-axis up the time-axis remote to where the mass is. So, what we see is mass mediating the relative strengths of the forces acting on space and time. The presence of mass changes the balance of the forces [tx] and [xt] in favour of [tx]. That has the direct effect of slowing the progression of the x-axis in the time direction – up the time-axis. An indirect consequence is the contraction of the x-axis – the contraction of space – the *gravitational effect*.

Comparing TRADD with the standard model

When we stop to think about this, the alternate perspectives in the standard model and the TRADD model become clear.

In the standard model, mass bends spacetime, and thereafter, bent spacetime manifests the gravitational effect. The gravitational effect operates in a static (unmoving) spacetime. It is the objects *in* space (objects having mass), that are mobile, navigating the shortest path through space warped by the presence of mass. It is the bent spacetime that creates the gravitational effect measured as the gravitational field. It is also the bent spacetime that slows time and contracts lengths. A relationship is drawn between the strength of the *gravitational field* and the amount of time dilation and length contraction. And from this perspective, it is understandable that gravity is regarded as the cause.

In the TRADD model, mass mediates force, directly slowing time. Hence, time dilation is relative to the amount and concentration of mass. In this regard, the two models agree. When mass mediates force, that also causes space to contract. Contracting space is the gravitational effect. Here, space is not static. As it contracts, it moves. And as it moves, objects in that space are pulled along with it. Any resistance to being pulled along as space contracts will be felt as a force. We call that force, gravity. Again, in this regard, the models agree. If we measure that force at different locations, the closer we get the point mass, the stronger the force will be. We could use such a collection of measurements to describe the gravitational field about the point mass. But length contraction is independent of the contracting space. Instead, it is related to the relative positions up the time-axis of the length being measured and the observer. The further away in time the measured length is, the more contracted it appears to be. This phenomenon results from the relative expansion rates of space of n-metres per year, but when a year passes more quickly for the observer than for the space wherein the measured length is located. Where the location of the length being measured is lower down the time-axis, in relative terms, the space there expands more slowly, so that lengths there are comparatively shorter because they have not had time to lengthen. But because we associate time dilation with the gravitational field, and because objects lower down the time-axis are indeed lower due to the presence of mass, the standard model easily associates length contraction with the gravitational field.

It should be clear now that when talking about gravity in the standard model or in the TRADD model, we are talking about the same thing. Both models allow for graded measurements of force about a point mass that describe a gravitational field. Both models relate time dilation and length contraction to the presence of mass. The standard model calibrates time dilation and length contraction relative to the strength of the gravitational field. It then finds that time dilates and lengths contract in a gravitational field. Which is true. But we then somehow believe that gravity is the cause. That gravity, or the *gravitational field* is responsible for time slowing down and lengths getting shorter.

TRADD agrees with the measured gravitational field but does not attribute time dilation to it. Even as the gravitational field can be used to calibrate it, TRADD does not find the field is causal. TRADD finds that the mediation of force by mass causes both time dilation and gravity. And TRADD finds that the degree of time dilation governs the degree of length contraction. The measurements are the same. But what is happening draws different perspectives depending on which model you look at.

Further, where the standard model attributes the curvature of spacetime to mass, it does this through a failure to acknowledge the contraction of space about a point mass. That contraction sets up a kind of *flow* of space – movement. By regarding space as static, the standard model interprets this movement as a curvature of the static space – indeed as the curvature of a static spacetime. In this context, not even the flow of time is acknowledged. Both time and space as components of spacetime, are static, and curve (warp) in the presence of mass. While TRADD does acknowledge curved space, it does not acknowledge curved time, and hence it does not acknowledge curved spacetime – only curved space. But whereas the standard model finds mass is directly responsible for curved spacetime, and hence curved space, TRADD finds mass is only directly responsible for the slowing down of time, (not curved time) – for the slower progression of the x-axis up the time-axis at the point mass. Curved space in TRADD results from the x-axis being bent in the time direction due to the x-axis lagging in time about a point mass. Under TRADD, gravity does not constitute a bending of space but rather a contraction of space accompanied by the movement of it due to the contraction. Hence, curved space under TRADD is not that same curved space acknowledged by the standard model. Curved space under TRADD cannot be compared with the curved space in the curved spacetime under the standard model where it is associated with gravity. Curved space under TRADD is solely associated with time differences between locations along the x-axis. Extended to a 3-dimensional universe, curved space under TRADD is associated with time differences between different locations in space. Under the standard model however, because we are unable to measure space bent in the time-direction, we only measure the associated time difference, call it time dilation, and associate it to the gravitational field.

Let's take a closer look at how this all works.

Mass – falling through spacetime.

Up until now, we said that a point mass slows the progression of the x-axis upward through the time-axis. We showed that the larger the point mass, the slower the x-axis moves upward through time. But that presents only half of what is really going on. We know from the standard model that if we take two clocks and set them to the same time in a location at sea level, then carry one of those clocks to the top of a mountain, it will progressively tick faster. When we bring it back down the mountain to sea level, the time on the two clocks will be different. They will be ticking at the same speed again, but the clock left at sea level will show an earlier time than the clock that went up the mountain and back again. That's because the clock that went up the mountain sped up so that its time got ahead of the clock left at sea level. The reason given is that the closer a clock is to the point mass, the Earth in this case, the slower time passes. And the further away a clock is from the point mass, up the mountain in this case, the faster time passes.

That means the proximity to a point mass does not simply produce an earlier or later time but produces a *slower* or *faster* time – a different *rate*! Earlier, we showed a point mass slowing the progression of the x-axis up the time-axis as a time *difference* – an earlier and later time. Now we need to show it as a continually increasing time difference – we need to show different time *rates*.

Reintroducing a point mass

In the treatment of time and a point mass below, we temporarily suspend the effects of x-axis *contraction* due to the point mass, that is, the effect of 'gravity' on the x-axis – what we call the *material contraction*. We do this because we want to illustrate the effect a point mass has on time, which occurs independently of the effect a point mass has on the x-axis — a contraction. Both phenomena continue to give effect to the 'length' of the x-axis. But the effect the point mass has on contraction is a local effect that dissipates quickly with distance from the point mass. The effect the point mass has on time, however, increases with distance from the point mass. It is this effect that we want to understand more completely.

At the point mass, there is a contraction in the x-axis – space contracts. But as we move away from the point mass, the contraction slows. Because the x-axis continues to move up through the time-axis, the x-axis expands as it does. Close to the point mass, the *expansion* due to the x-axis moving forward in time is less than the *contraction* due to the presence of the point mass. So here the net result is a contraction; here the 'gravitational' effect dominates; there is 'gravity' and bodies are pulled toward the point mass. Moving away from the point mass, there comes a point where the *contraction* balances the *expansion*. Move further away, and the expansion exceeds any remaining effect on the 'length' of the x-axis produced by the contraction – the net result is an expansion and bodies are pushed away from the point mass (by the expanding space).

To summarise, space contracts in two ways: materially and relatively. First, *locally*, time slowing due to the presence of mass rebalances forces to produce a material *contraction* in the x-axis – the 'gravitational' effect. Second, time slowing due to the presence of mass, produces a slower x-axis expansion with progress up the time-axis relative to the expansion of the x-axis remote to the mass where it travels more quickly up the time-axis. The slower expansion about the point mass amounts to a contraction *relative* to the faster expansion remote to the point mass.

The x-axis *material contraction,* the 'gravitational' effect, reduces with increasing distance from the point mass along the x-axis.

The x-axis *relative contraction* between separated points along the x-axis *increases* with their separating distance along the time-*axis*.

In the chart below, we see a point mass at x=-1 where it has slowed the progression of the x-axis up the time-axis. This means that along the x-axis at the point mass, time is progressing slower than at distances to the left and right of the point mass. The further away we get from the point mass the faster a clock would tick, until we are so far away (x=-2 and x=0), that the influence of the point mass is so small as to be negligible. Also, where time progresses slower at the point mass, the x-axis, which expands at n-metres per year, expands slower than further away where time progresses more quickly. Whilst always expanding at n-metres per year, because it takes longer for a year to elapse at the point mass than further away from it, the expansion by n-metres per year takes longer at the point mass than it would further away from it. Viewed by an observer to the right of the time-axis, the x-axis has shrunk in the vicinity of the point mass – lengths have contracted.

Note that the vertical distance from the x-axis at the point mass to the x-axis remote to the point mass is approximately 0.2 years. That means an observer remote to the point mass would measure the time at the point mass to be about 0.2 years behind. It's the same as an observer at the top of the mountain with the clock carried up from sea level. If the observer were able to see the clock left behind at sea level, the clock at sea level would read an earlier time than the clock carried to the top of the mountain.

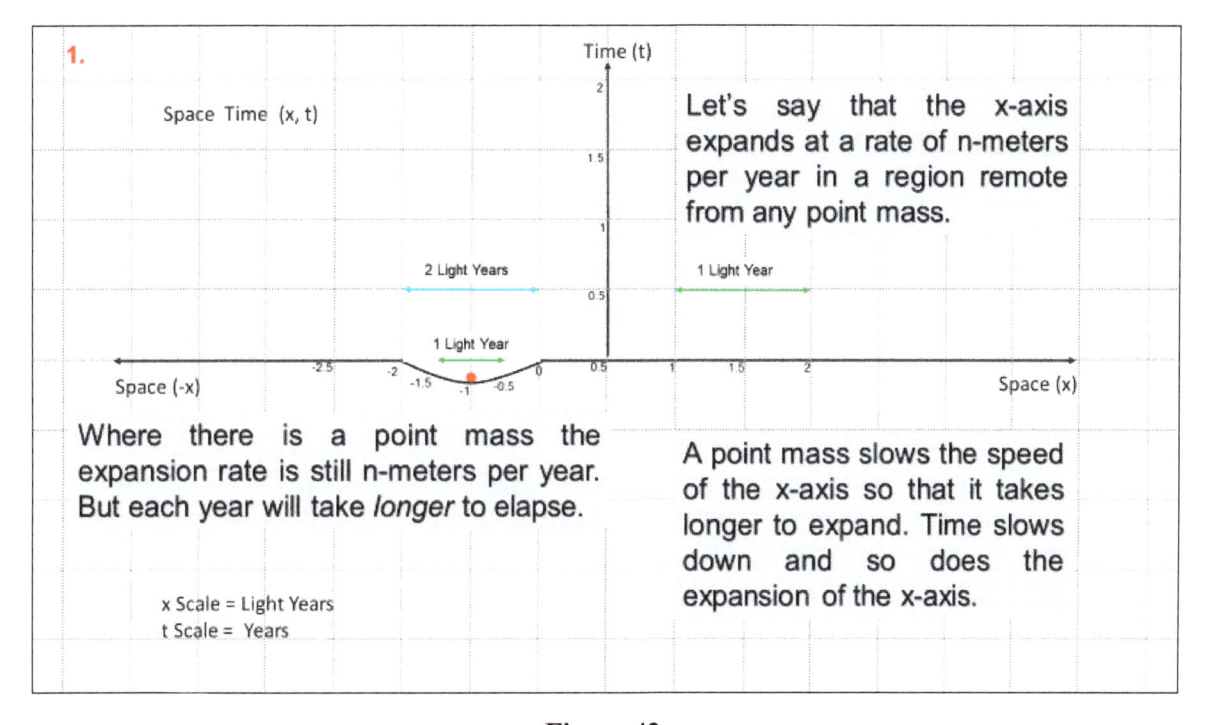

Figure 42.

How time slows down

In chart 2 below, we see how the x-axis has progressed upward through the time-axis from t=0 marked by the orange line. We have not recalibrated the time-axis. The x-axis is now at t=0.5. At the point mass along the x-axis, the vertical distance from the point mass to the x-axis remote to the point mass has increased. That's because at the point mass, the x-axis is moving upward at a slower pace, as compared to the pace of the x-axis remote to it.

Note that the vertical distance from the x-axis at the point mass to the x-axis remote to the point mass is approximately 0.3 years. That means that an observer remote to the point mass would measure the time at the point mass to be about 0.3 years behind. Last time we checked, it was around 0.2 years behind.

We also note that because *at* the point mass, progress upward through the time-axis is slower, a year will take longer to elapse. The x-axis is expanding at a constant rate of n-metres per *year*, but the elapse time of a *year* takes longer at the point mass. Lengths at and around the point mass become progressively shorter when compared with lengths remote to the point mass. The expansion of the x-axis in the vicinity of the point mass slows more and more as compared with the expansion of the x-axis remote to the point mass, due to progressively shorter lengths expanding at n-metres per year. 1 light-year remote to the point mass (to the right) is measured as 3 light-years in the vicinity of the point mass (to the left). Therefore, 1 light-year about the point mass appears to have contracted to a distance equivalent to one third of a light-year when measured by a remote observer.

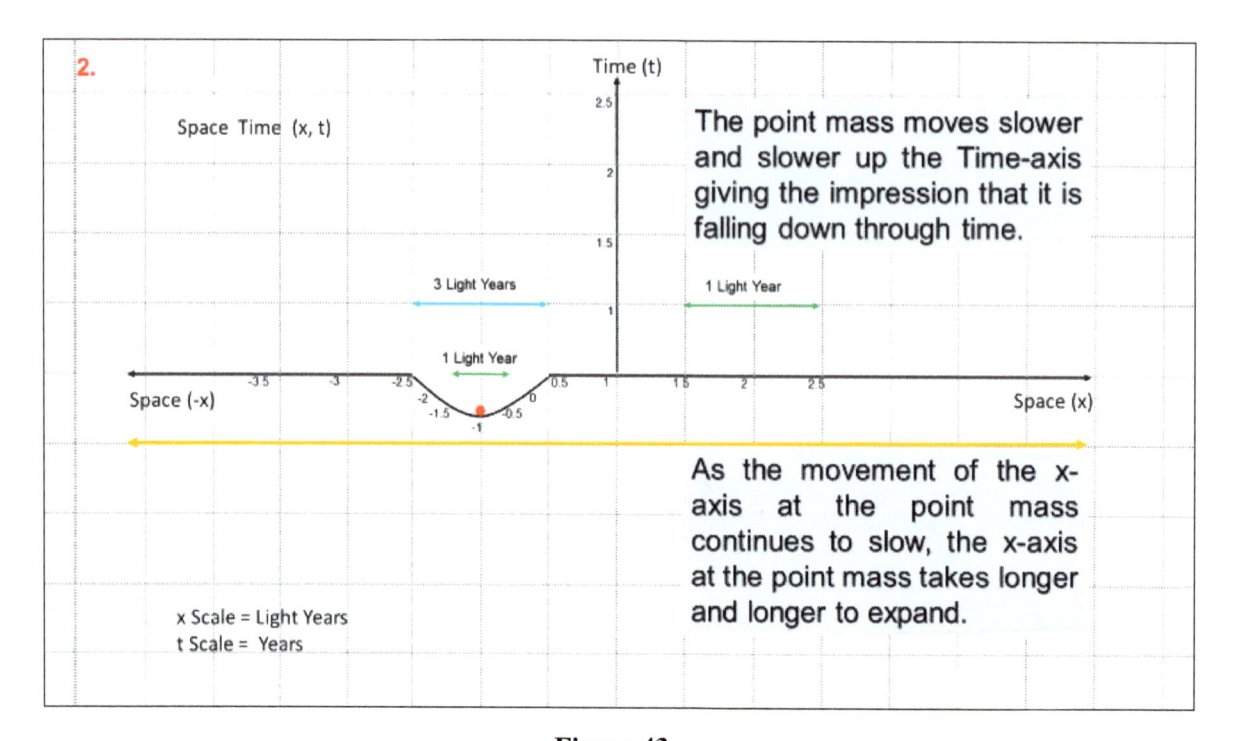

Figure 43.

Falling through time

In chart 3 below, we see how the x-axis has progressed upward through the time-axis from t=0 marked by the orange line. We have not recalibrated the time-axis. The x-axis is now at t=1. At the point mass along the x-axis, the vertical distance from the point mass to the x-axis remote to the point mass has increased. This happens because at the point mass, the x-axis is moving upward at a slower and slower pace, as compared to the pace of the x-axis remote to it.

Note that the vertical distance from the x-axis at the point mass to the x-axis remote to the point mass is approximately 0.5 years. So, an observer remote to the point mass would measure the time at the point mass to be about half a year behind. Last time we checked, it was around 0.3 years behind. Again, we see, that as the x-axis progresses upward through the time-axis, the x-axis at the point mass is getting further and further behind – in time. It is as if the point mass is falling through time.

Again, *at* the point mass, progress upward through the time-axis is slower, so a year takes longer to elapse. With the x-axis expanding at a constant rate of n-metres per *year*, but the lengths getting smaller and smaller as compared with lengths remote to the point mass, the expansion of the x-axis in the vicinity of the point mass continually slows down as compared with the expansion of the x-axis remote to the point mass. Now, 1 light-year remote to the point mass (to the right) is measured as 4 light-years in the vicinity of the point mass (to the left). Now a distance of 1 light-year about the point mass appears to have contracted to a distance equivalent to one quarter of a light-year when measured by a remote observer. Space (the x-axis) appears *compressed* – contracted – in the vicinity of the point mass as compared with space (the x-axis) remote to the point mass. Not only does it appear contracted – but it also appears to be *contracting* over time.

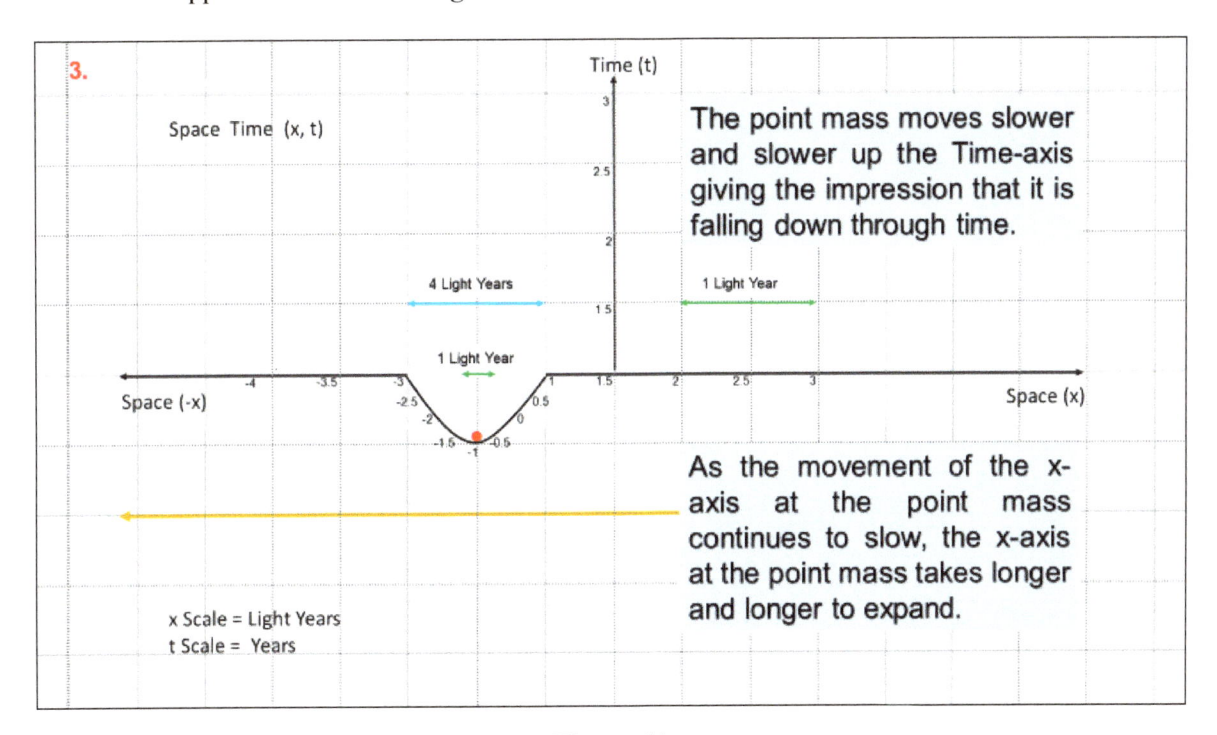

Figure 44.

Falling through space

In chart 4 below, we see how the x-axis has progressed further upward through the time-axis from t=0 marked by the orange line. We have not recalibrated the time-axis. The x-axis is now at t=1.5. At the point mass along the x-axis, the vertical distance from the point mass to the x-axis remote to the point mass has increased. That's because at the point mass, the x-axis is moving upward at an ever-slowing pace, as compared to the pace of the x-axis remote to it.

Note that the vertical distance from the x-axis at the point mass to the x-axis remote to the point mass is approximately 1 year. That means an observer remote to the point mass would measure the time at the point mass to be about 1 year behind. Last time we checked, it was around 0.5 years behind. So again, we see, that as the x-axis progresses upward through the time-axis, the x-axis at the point mass is getting further and further behind – in time. The point mass appears to be falling through time.

But hang on!

Something else is becoming apparent.

From the perspective of an observer remote to the point mass, the x-axis at the point mass appears to be *falling* through time. We say that, because as the x-axis progresses upward through the time-axis, the time **difference** between the point mass and an observer remote to the point mass **continues to increase**. But if we were paying attention, we also said that the x-axis in the vicinity of the point mass is expanding at a slower rate than at a location remote to the point mass. That means that the x-axis

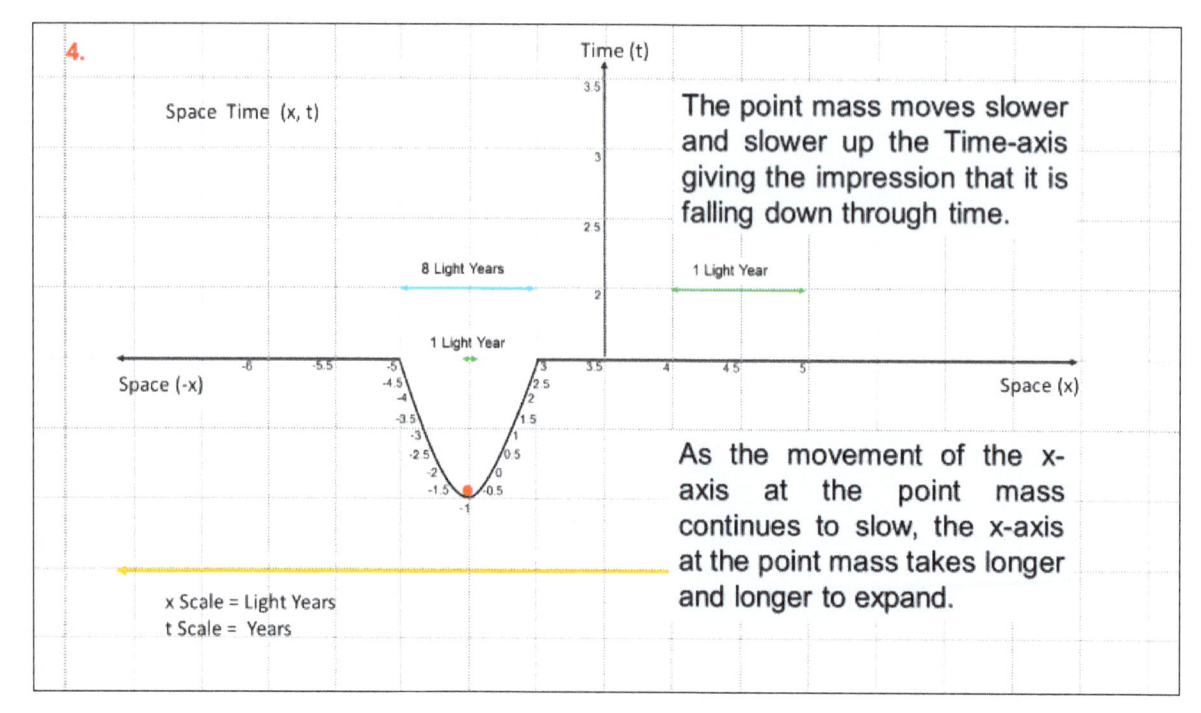

Figure 45.

remote to the point mass is expanding faster than at the point mass. And *that* means that for a remote observer, the *distance to* the point mass, the ***distance difference***, well, it ***continues to increase***. And the further away from the point mass you are, the faster the distance to the point mass increases – the point mass is *accelerating* away! In other words, the point mass is getting further and further away – in space. From a remote observer's perspective then, the point mass appears to be *falling* through time but also *falling* through space! And at an accelerating rate.

Falling through spacetime

If we regard our chart with the x-axis and time-axis as a map of space and time and if we then further consider the inclusion of the y-axis and the z-axis, we could name that four-dimensional construct: spacetime.

Then, what we have shown is, from an observer's perspective, a point mass appears to fall through spacetime when observed from a location remote to the point mass; sufficiently remote that the influence of the point mass is negligible. We see that no matter your distance from the point mass, the distance will increase with time as space expands. We see that the further away from the point mass you are, the faster the point mass appears to fall through spacetime. We also see that the bigger the point mass, the faster the point mass falls through spacetime.

So, we see that how fast a point mass falls through spacetime depends on the time difference, the size of the point mass and the distance separating the point mass from the observer. And as we have seen, time is always measured locally – it is the speed with which a point on the x-axis moves upward through the time-axis.

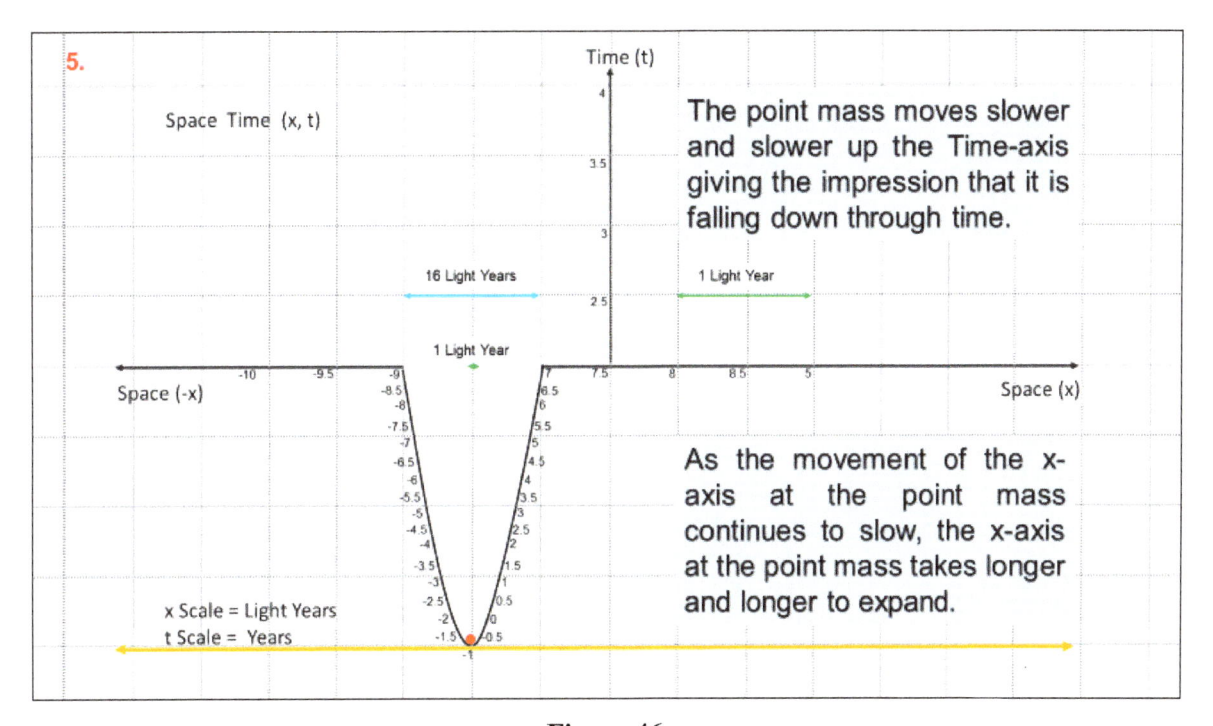

Figure 46.

Reconstructing the criterion

Let's look, *yet again,* at what we have done.

Chart 1. We reintroduced a point mass at -1 on the x-axis.

Chart 2. We show how time slows down.

Chart 3. We show how mass falls through time.

Chart 4. We show how mass falls through space.

Chart 5. We show how mass falls through spacetime.

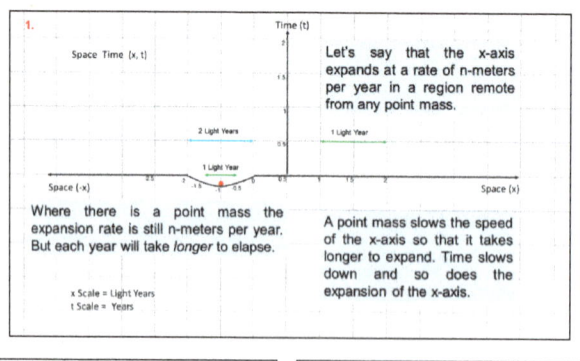

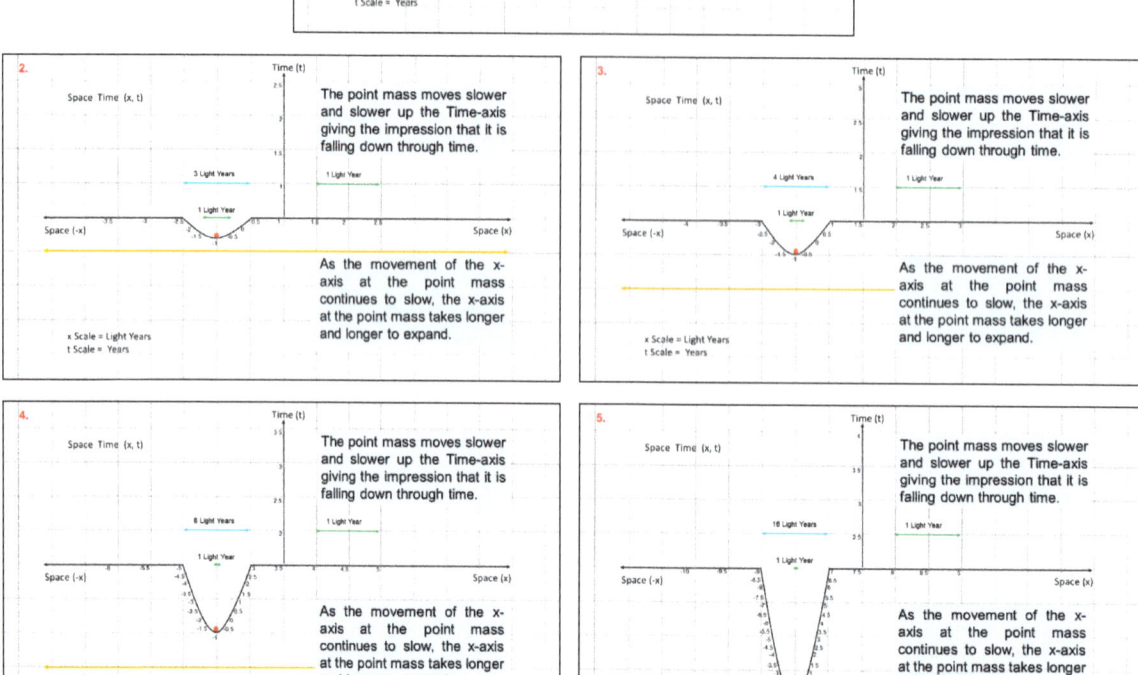

Figure 47.

Flattening time

Up until now, in the charts we have used to depict what is going on when a point mass is present on the x-axis, the x-axis depicts a direction in space, or in spacetime, if you like.

But when we take measurements in space from a position that we observe from, we don't think of the x-axis as being curved in the vicinity of a point mass. Whilst that is the outcome from general relativity theory in the standard model, when we say, look at the Sun from Earth, we pretty much see it along a straight line.

In other words, when we look at the Sun from Earth, we believe we can draw a straight line from the surface of the Earth to the surface of the Sun. And when we measure the length of that straight line, we believe we get the distance of the Sun from the Earth.

When we look out into space, we do not see that it is curved in the way that we might see an undulating road before us dip and rise as it follows the contours of the land. Our standard model tells us it is curved – because spacetime is curved – and that has been confirmed by measurements that uphold the theory of general relativity. Light's path bends toward the Sun due to the Sun's gravitational field. We measure that curvature in space – but not in time. Why is it that we can measure space as curved but not time? How do you bend time? Surely, if you can bend spacetime then you can bend time. We ask this because in the standard model, it is spacetime that is curved. But in TRADD, there is no curvature of time, only of space. Perhaps, that's significant.

When space is bent in the time direction

When we look out into the universe, we see things because the light emitted from or reflected by an object reaches our eyes or our instruments. We also have this 'rule' that light travels in straight lines. And so, there's always a straight line of sight between us and the object we are looking at. Well, that's unless the straight line got bent passing a point mass. Then we see the straight line is bent in three-dimensional space – in some combination of the three spatial directions x, y, and z.

The standard model tells us that light travels on a path through spacetime we call a geodesic.

If you draw the shortest possible line between two points on the surface of a sphere, that line is a geodesic. So, even as we can see the line is curved along the surface of the sphere – it remains the shortest distance between the two points – a pseudo straight line.

Another way to think about it, is to take a sheet of paper, lay it down flat. Draw a straight line on it. Look at your line. It's a straight line. Now lift one corner of the sheet so that the sheet is bent. The line you drew is bent, too – bent because it just follows the *curve* in the sheet created when you lifted the corner. So, it's not a straight line anymore – it's a geodesic!

A geodesic is what 'passes' for a straight line in spacetime. It defines the shortest path between two locations through curved spacetime. But unlike when we look at a geodesic on the surface of a sphere, a geodesic through spacetime can appear to us as a straight line. We don't always *see* the curvatures in space. And that's because, as we have seen, space, the x-axis, is sometimes bent in the time direction. And with our senses, we are unable to see it when space is bent in the time direction. We *can* detect when lines appear bent in one or more of the three spatial directions – but not when it's *bent* in time.

Instead, in the context of gravity, the standard model defers to a concept called time dilation – *more about that later*. The standard model does not say that when spacetime is bent, that time is consequently bent. It instead says that time slows down, which connotes a flow, and that spacetime is curved, which does not connote a flow. So we are left with the same phenomenon, gravity, slowing the progression of time, as well as bending spacetime. Let's be very clear about this. It is not space that is curved by gravity in the standard model. Spacetime is curved. So on one hand, we have curved spacetime from which we do conclude that space is curved. On the other hand, we have curved spacetime from which we do *not* conclude that time is curved. Instead, time slows down. Hmmm …

Of course, TRADD deals with the notion of curved spacetime a little differently. TRADD acknowledges that space can be bent in the time-direction, but that this effect is only indirectly caused by mass, by gravity. TRADD says that spacetime isn't curved at all due to gravity. Instead, space contracts. Space moves. It also moves about due to the general expansion of space as the x-axis progresses up the time-axis. But because in the standard model we regard space as static, our measurements relating to space flows are interpreted as measurements relating to space curvatures. Now we are about to understand why this interpretation is so problematic for the TRADD model. And the problem here has to do with time. Under the standard model, space is static, but time *flows*. And that makes for a difficult task. Reconciling the curvature of spacetime with the idea that time flows while space which is static, curves.

Straightening out the curves

In the chart 1 below, we see that the x-axis is bent in the time direction – well, in the negative direction of time to be precise. We have already shown that this occurs because the point mass slows the progression of the x-axis upward in the time-direction, giving the appearance of the point mass falling through time. Indeed, from the perspective of an observer remote to the point mass, it is falling at an accelerating rate, faster and faster – in the time direction. We can measure this using clocks.

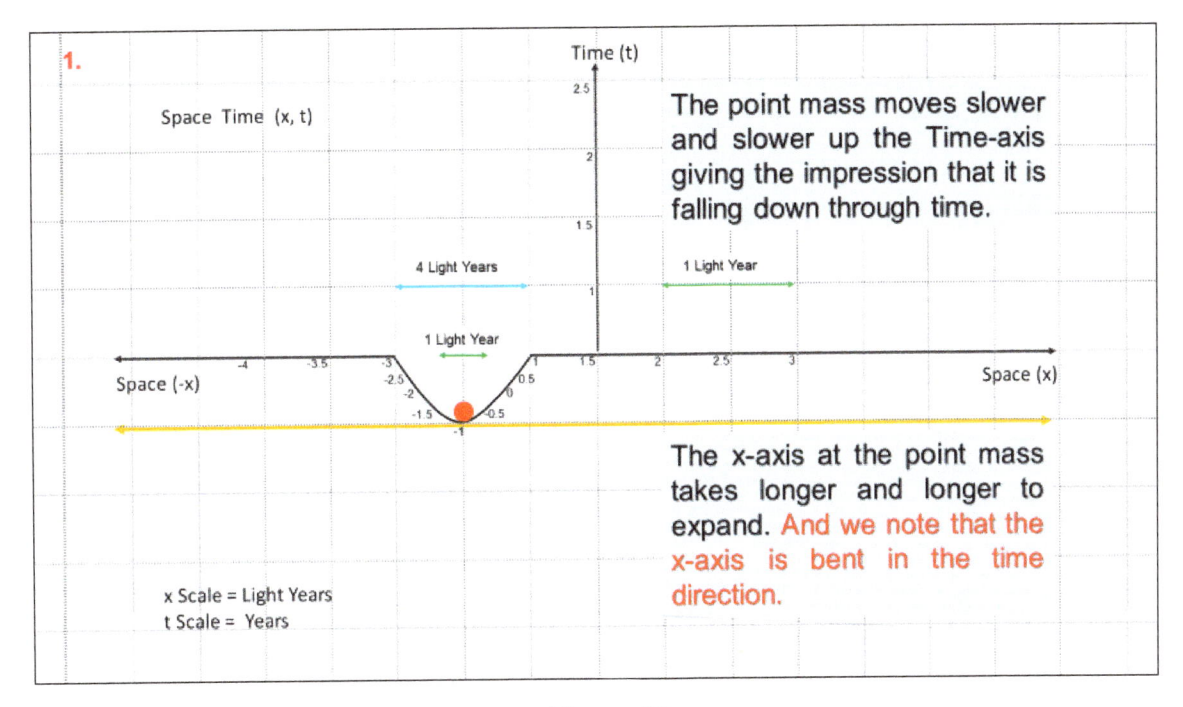

Figure 48.

We want to see what it will look like if we straighten out the bent section of the x-axis. Which is how space would appear to us and our instruments when we observe it.

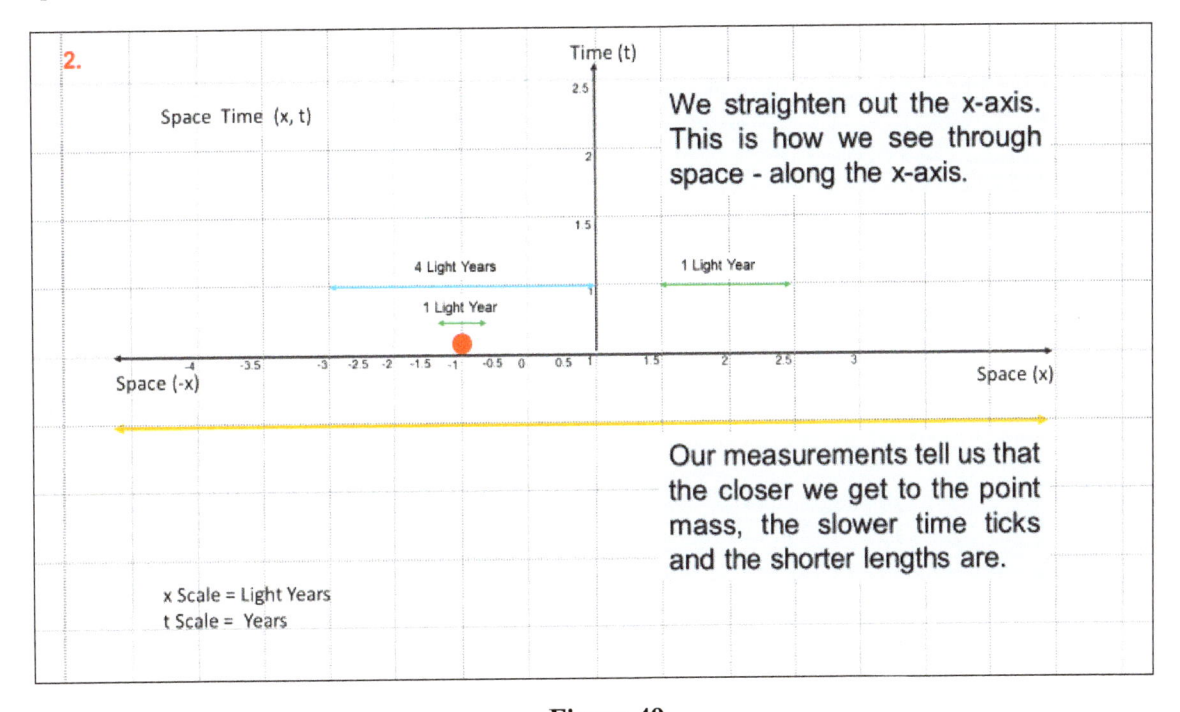

Figure 49.

99

When we look through space, we are unaware of the curvature of the x-axis. Instead, we view the x-axis as being a straight line. To us, the x-axis is a line drawn on a piece of paper lying flat on the table. But that does not change what is happening at the point mass.

When we take measurements at the point mass from a position on the x-axis remote to the point mass, we note that time slows down, and lengths get shorter. We correctly attribute these phenomena to the presence of the point mass.

Let's think again about how time slows down.

We have shown that time slows down progressively more and more the closer we get to the point mass on the x-axis. We have shown that this slowing down of time results because at the point mass, the x-axis moves more slowly upward through the time-axis. And the x-axis progressively moves faster up the time-axis the further away from the point mass you get.

Let's also think again about how lengths contract.

We said that the x-axis stretches at a rate of n-metres per year and that a year takes longer to pass at the point mass than remote to it. Hence, stretching the x-axis takes longer at the point mass than remote to it. Also, at the point mass, lengths are getting comparatively shorter and shorter as time passes, as compared with lengths remote to the point mass. Hence, shorter and shorter lengths are expanding at a slower and slower rate. The further away you get from the point mass, the faster the x-axis stretches as compared to how fast it stretches at the point mass. But whether at the point mass or remote to it, the expansion of the x-axis remains always at the rate of n-metres per year.

So, what's the problem here?

The problem we introduce when straightening out the x-axis, flattening it, is that in doing so, we needed to compress the time-axis. We straightened the x-axis at and around the point mass. But time values at and around the point mass that were less than time values remote to it, now all have the same value. In other words, in straightening out the x-axis, we have distorted time. Where before there was a measurable relationship between time and the expansion of space (expansion along the x-axis), that relationship is lost when we *artificially* straighten out the x-axis because doing *that* distorts the time values. Before, at any point along the x-axis, space was expanding at the rate of n-metres per year. And that was true so long as the progression of the x-axis up the time-axis varied according to its proximity to the point mass. But having flattened the x-axis, its progression upward along the time-axis is uniform. It's the same speed everywhere along the x-axis – even at the point mass. And that's just not how things work.

This is what happens in the standard model. It takes the x-axis as being straight (with respect to time). And so, whilst general relativity can account for the bending of spacetime due to a point-mass, it does not account for this bending as being fundamentally in the time direction. Instead, it starts out with a spacetime grid, notes the bending of space due to a point mass, and finds that time consequently gets bent, (dilated), because it forms part of the spacetime grid. It says that space cannot bend without also bending time because time is part of the fabric of spacetime. Bending space bends the fabric, which bends time. Or bending space bends the fabric, which *dilates* time.

Our new TRADD model of the universe agrees with the observations and measurements in the standard model. Where the TRADD model differs is in its perspective. It finds that a point mass, sets up a flow in space that the standard model will interpret as a bending of spacetime. In the absence of a flowing space, TRADD shows mass bends space in the time direction. It shows that this *bending* of space (the x-axis, extensible to the y-axis and the z-axis) is what we conclude, not because space is being pushed around and bent in the way one might bend a length of wire, but because at a point mass, space lags in time, bending the x-axis in the time direction. Further, the slowing down of time and contraction of lengths occur as a natural consequence of space lagging in time at the point mass. Each of these three phenomena that occur due to the presence of a point mass in the standard model - the bending of spacetime, time dilation and length contraction – are shown to result from a point mass lagging in time. Or, more specifically, lagging the progression of space *through* time – as we have shown using the progression of the x-axis up through the time-axis. If we repeat that with the y-axis and the z-axis, then what we have is the progression of observable three-dimensional space lagging through time.

So it is, that the standard model measures in ever closer proximity to a point mass, the slowing down of time, the progressive bending of spacetime and the ongoing contraction of lengths. It presents these phenomena because a point mass is present. But beyond the point mass being present, affords no further explanation.

In the standard model, we don't know *how* or *why* 'matter bends space'.

In the TRADD model, we do.

I'm not saying that the TRADD model is correct. What I am saying, is that it offers a perspective on what might be going on that remains consistent with our measurements and observations under the standard model, but also, that just might make sense.

So, is gravity a force or not?

What then accounts for our *experience* of gravity as an attractive force?

We have described how gravity comes about due to the contraction of space, not unlike a stretched rubber band being unstretched. Thereby, it behaves as space leaking out of the universe – out through the point mass, such as the Earth. And as space leaks out, it moves.

Jumping up off the ground is akin to swimming against a strong current in a river. You can make some headway with respect to the riverbed but the moment you stop swimming you are again swept along by the current. If you stood firm in the river, feeling the strength of the current is like how we feel gravity. Getting swept along by the current, that is like falling due to gravity. Space has currents that flow in the direction of matter. With water leaking out through a hole in a bucket, the closer the water is to the hole, the faster it flows. With space flowing toward matter, the closer the space is to the matter, the faster it flows.

We say evidence for the expansion of space can be seen when observing two galaxies accelerating away from one another. Our TRADD model says that the space between them expands. If we say that as the universe (space?) expands between galaxies thus pushing them apart, does that not imply pressure? Under TRADD, space gives the appearance that it does not compress. Create more of it, and it will push existing space out of the way. And that *new* space, pushing the existing space out of the way to make room for itself, produces a current as well. The *pushing* implies a force. And that space does not compress, implies *pressure*. The existing space needs to get out of the way to make room for the new space. It must move!

It doesn't matter how you want to think about it. Take an elastic band and draw three dots on it equal distances apart. Now stretch the elastic band. The middle dot might stay where it is relative to something that catches your eye, but the dot's either side of it – they must move.

When you stand on the surface of the Earth, the space around you is contracting relative to the space further away from the Earth's centre due to the mass of the Earth lagging the planet's progression through time. The Earth is falling through spacetime due to its mass. And you fall through spacetime with it. But as described, falling through spacetime is relative to a remote observer. Locally, the effect is space contracting, creating movement of space toward the point mass. So, jump up in the air and you will be pushed back down – by the space current.

This, I believe, was the essence of Einstein's insight.

If you were falling toward the Earth from a height – and ignoring air resistance and any visibility of objects about you – you would have no sense of motion. Nor would you *feel* the force of gravity – you would not feel your own weight. Our new model says you would simply be floating along with the current, the space current.

But just because you are 'falling' through spacetime does not mean you are not also moving through time. *Falling* is completely relative. It is relative to a location in spacetime remote to the point mass. We have shown that a point mass can move forward through time even as, in a relative sense, it appears to fall *backwards* through spacetime. It is a quirk of nature, that a slower progression through time does not change the expansion rate of n-metres per year. That is because a year will eventually elapse – and once elapsed, space will have expanded by n-metres. In a region of space where a year takes longer to elapse, space takes longer to expand – but only as compared with a region where a year elapses more quickly.

You can see now that galaxies in an expanding universe can act as point masses contracting the space around them. So, when galaxies are near enough – when their point masses are close enough on the x-axis, they *fall* toward one another. Where they are far enough apart – remote to each other's point mass, then the space between them expands at an ever-increasing rate; and as observed, they accelerate away from one another.

So, where the *expansion* of space between galaxies is greater than the *contraction* of space in the proximity of their point masses, they move apart – otherwise they move together. These two mechanisms – space contracting at a point mass and the expansion of space as it moves forward in time, act to pull galaxies together or push them apart.

A region's slower progression through time is seen as *a region of contracting space* from the perspective of a region progressing faster through time. From the perspective of a region progressing slower through time, the region progressing faster through time is seen as *a region of expanding space*.

So, is gravity a force?

It is.

It is the force of space!

But we don't measure the *pressure* in the universe. We don't measure the *force* exerted when existing space is pushed out of the way to make room for new space – when space stretches. We don't measure the local contraction of space due to a point mass.

We see them as two different things. Our universe is expanding – that's one thing. Lengths contract in a gravitational field – that's another thing. We don't *see* the space currents. We only see their effect on the stuff *we can see* and measure.

We are happy to acknowledge that spacetime – our universe – is expanding. We are happy to go with the idea of the 'Big Bang' theory that requires a single burst of energy to get the whole process going. But we then get stuck with the idea of 'dark energy'. You see, with the physics we have, acceleration requires the consistent application of a force. If galaxies are accelerating away from one another, if therefore our universe is expanding at an accelerating rate – where is the force responsible for that acceleration? Where does the energy come from to give presence to that force? We don't know – so for want of a better term, we made one up. We call it dark energy. Dark, because we can't see it.

Space exerts force when it expands *and* when it contracts. When it expands it pushes everything outward. When it contracts it pulls everything inward. We can feel the force it exerts. The force of space.

But when we measure it – we just don't call it that – and that's because we don't see that space moves – we don't see space at all – we only see things *in* space.

What is light?

We need to find a different way to describe light. One that fits with our different description of space and gravity. We can start by finding a way to describe what we already know about light from the measurements we've taken. We know light travels at a constant speed. We know it manifests as a particle and as a wave. No one knows why it travels at the speed it does. No one understands why, when its speed is measured, we always get the same value regardless of the motion of the observer. No one knows why it presents as a particle and as a wave. Our challenge is to find a description of light that explains how and why it presents as a particle and a wave, and how and why it constantly travels at the speed it does.

The speed of light

Light travels at 299,792,458 meters per second. That's not only a big number – it's very fast! It is also regarded as a universal constant denoted by the symbol *C*. Whenever the speed of light is measured, we get the same value – it's constant. We get the same value whether we are stationary or moving toward or away from the light.

A light-year

A light-year is the distance light travels in 1 year: 9.461×10^{15} metres. That's 9,461,000,000,000,000 metres, or 9,461,000,000,000 kilometres (nine thousand, four hundred and sixty-one billion kilometres). The symbol used to denote 1 light-year is *ly*.

Something doesn't add up.

Let's start with a 'thought experiment'.

Let's have a stretch of space from left to right marked off in half light-years and place a light source at the left end of the stretch. We will set up the light source to emit photons from a time we will call time zero. But not just any emission of photons. We'll emit the photons only once every day and emit an increasing number of photons each successive day, spaced a second apart. The number of photons will equal the count of the number of days since time zero. For example, at the end of the day, after 1 year, 365 photons will be emitted, one second apart. That way, anyone seeing the light to the right of the light source can know how many days after time zero the photons were omitted, simply by counting the number of photons in the emission burst.

Also, we fly a traveller from left to right at half the light-speed, passing the light source at time zero. After three years, the traveller will be 1.5 light-years to the right of the light source.

Now here's the thing.

Having travelled 1.5 light-years, the traveller looks back and measures the emission burst. The traveller is interested in two measurements. What is the approach speed of the photons? How many days have the photons been travelling?

The traveller determines that the approach speed is the speed of light and reasons that given the traveller's speed away from the photons is half light-speed, the speed of the photons must be light-speed plus the half light-speed the traveller is moving at. That sums to 1.5 light-speed for the photons.

But photons are only supposed to travel at light-speed – 1 light-year per year. Neither slower nor faster.

But it's ok.

The emission burst will have encoded in it the number of days the photons have been travelling. All the traveller needs to do is to subtract the number of photons in the emission burst from the number of days since time zero. So, subtract the number of photons from the number of days in three years. If the photons are indeed travelling at 1.5 light-speed, then the number of days the photons have been travelling ought to tally to 1 year. Then the photons would have travelled 1.5 light-years in the year they were travelling.

Now it turns out that the photons had indeed only been travelling for 1 year. They left the light source 2 years after the traveller passed it. The traveller was 1 light-year away when the photons were emitted. In the year taken by the traveller to travel the half light-year to 1.5 light-years from the source, the photons had travelled the full 1.5 light-years from the source to the traveller. The photons had covered the distance of 1.5 light-years during their 1 year of travel. And *that* – surprised the traveller, because light is supposed to travel at light-speed – which is 1 light-year per year – not 1.5 light-years per year.

So, how is it that we are so stuck on the idea that light *travels* at a constant speed? One light-year per year, right? And how is it that we measure the approach speed of light to always be that speed, regardless of a traveller's speed toward or away from the light source, and even if the traveller is standing still?

Our thought experiment suggests that light can travel faster than light-speed. And stranger than that – light's speed *seems* to be dependent on the traveller's speed. Yet, the traveller's speed was chosen at random. Had we chosen one-quarter light-speed for the traveller, the approach speed of light to the traveller would also have been measured to be light-speed. Speed calculations for the photons would not come out to be light-speed as expected, but light-speed plus the speed of the traveller. In this case, the photons would have travelled 1.25 light-years during their year of travel.

That's not supposed to happen!

But hey, it's only a thought experiment.

Ridiculous, isn't it?

On one hand, we are talking about light-speed as a 'constant speed'. We are able to demonstrate that as an approach speed, it is indeed constant. It's what we measure. On the other hand, we regard light-speed as a *property* of light. We expect the distance light travels each year to be one light-year. Indeed, it is this outcome that we use to measure the distances between things in the universe – even the age of the universe itself. When we say that something is 50 light-years away, we also mean that light will take 50 years to traverse that distance. We know that light travels exactly 1 light-year, per year, every year. Right?

Well … maybe.

Didn't we just 'thought experiment' our way to finding that light travelled 1.5 light-years in 1 year?

Something just doesn't add up.

But don't worry – there is, as there should be, a logical explanation.

So … what's going on?

Spacetime

Let's start with the standard model: the idea that the universe is a spacetime continuum. We could call the three directions (x, y, and z) and denote time as (t). We could then draw a graph of what the plane in (x) and (t) would look like. We have x as the horizontal axis and t as the vertical axis. We can measure distance along the x-axis in light-years and measure time along the t-axis in years. We can scale the x-axis in units of half light-years and the t-axis in units of half years.

You've seen this before when we described space and gravity.

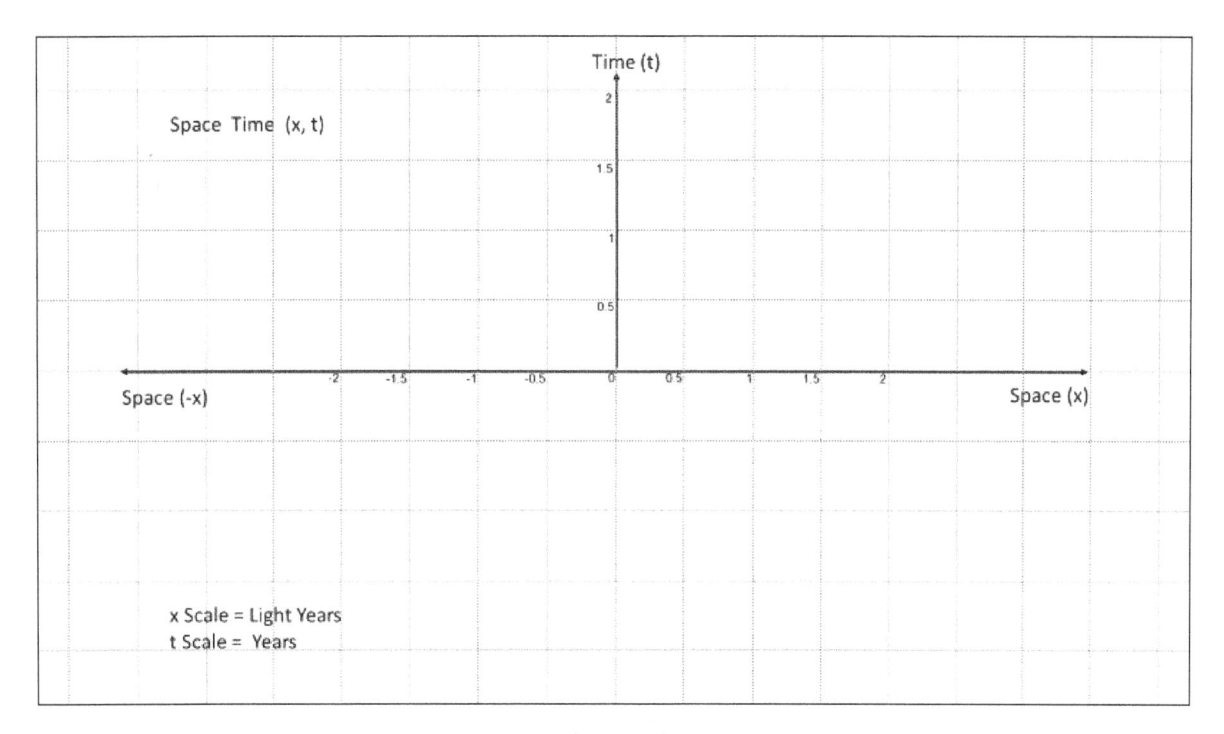

[Figure 50: Spacetime (x, t) coordinate graph with Time (t) on vertical axis and Space (x) on horizontal axis, both ranging from -2 to 2. x Scale = Light Years, t Scale = Years.]

Figure 50.

This is simply taking a two-dimensional slice of our four-dimensional spacetime. It will help us see what's going on.

How does light move?

We can start by figuring out how light moves through spacetime. Science already tells us that it moves in a straight line at a *constant* speed. We know that from observing light in experiments and taking measurements. The explanations about how light moves come directly from these measurements. But the explanations are not entirely satisfying.

If you sit in a parked car and I drive by in my red car at 100 km/hour, you could take measurements. You measure my approach speed as 100 km/hour and you measure the speed I travel away from you after I go past as 100 km/hour. If you were travelling on the freeway at 50 km/hour and I drove past with my car doing 100 km/hour, you could again take measurements. You measure my approach speed coming up behind you as 50 km/hour and the speed I travel away from you as 50 km/hour. If I was coming toward you because I was travelling in the opposite direction you could again take measurements. You would measure my approach speed as 150 km/hour and the speed I travel away from you as 150 km/hour.

But what if I travelled in the same way that light travelled, only my constant speed was 100 km/hour rather than light-speed? What if I had a blue car that could travel the way light does, but at 100 km/hour?

If you sit in a parked car and I drive by in my blue car at 100 km/hour, you could take measurements. You measure my approach speed as 100 km/hour and you measure the speed I travel away from you after I go past as 100 km/hour. If you were travelling on the freeway at 50 km/hour and I drove past with my blue car doing 100 km/hour, you could again take measurements. You measure my approach speed coming up behind you as 100 km/hour and the speed I travel away from you as 100 km/hour. If I was coming toward you because I was travelling in the opposite direction, you could again take measurements. You would measure my approach speed as 100 km/hour and the speed I travel away from you as 100 km/hour.

Do you see the problem?

If you took those measurements and used them to explain how my blue car travels, you can do that only because you can defend your measurements – not because it makes any sense. Anyone could repeat your experiment if I made my blue car available. They would all get the same results as you did. Everyone would say that my blue car travels at a constant speed of 100 km/hour and have the measurements to prove it.

But the constant speed of my blue car is not the same as if you were travelling in your car with the cruise-control set at 100 km/hour. You will say that you are travelling at a constant speed. But when you say *that* you know you are not saying the same thing as when you say that my blue car is travelling at a constant speed.

It's as though science hasn't found the language to explain how light moves. Science tells us light travels at a constant speed. What they really mean, though, is that whenever you measure how fast it's going, you always get the same value. It's not quite the same thing. I have heard it said that the speed of light is *a constant*. Now that makes sense when they add that it's going that speed no matter if you measure it whilst standing still, moving toward or away from it. And they do! Still, it's hardly compensation for a description that bears no resemblance to what we understand a constant speed to be.

It would have made as much sense if they said they didn't know – that its speed varies depending on the motion of the observer, but when measured it's always the same value. Remember our thought experiment? Hmmm.

But we do want to make sense of it. So, let's investigate!

We want to explore the path light takes through spacetime, starting from a point, 1 light-year away. We are interested in when we would see the light if we don't move and when we would see it if we did. When we move, we will travel very fast, through spacetime, in our spaceship. Using a scale of half years makes doing calculations easier because we know how far light would travel in that time. It travels 1 light-year in 1 year. So, in half a year it travels half a light-year. When light has travelled half a light-year, we know it has been travelling for half a year.

When we don't move

We have x as the horizontal axis and t as the vertical axis. We have a spaceship, the green marker, that we position at x=0 and t=0.

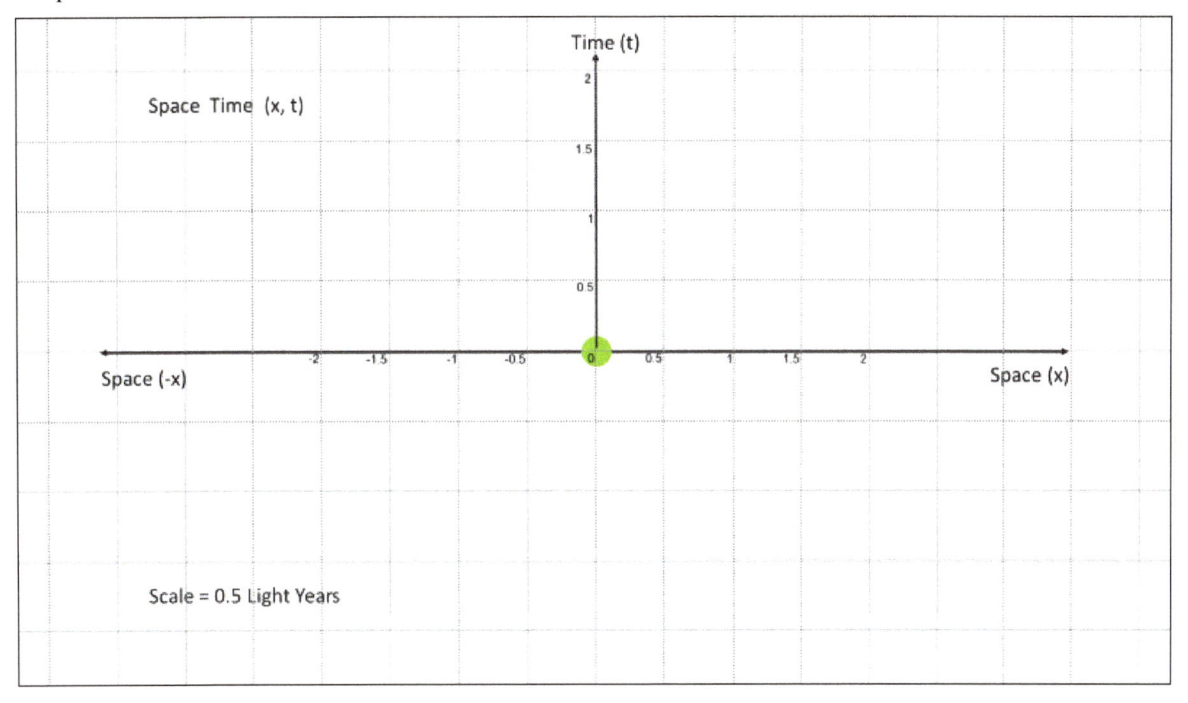

Figure 51.

Let's first look at what happens when we don't move. If our green marker is our spaceship, it would drift through time. Sitting on the launch pad does not change its position in space but it does change its position in spacetime. That's because it will move through time regardless of whether it's moving or standing still. Standing still for half a year would move it through spacetime to a point on the graph at x=0 and t=0.5

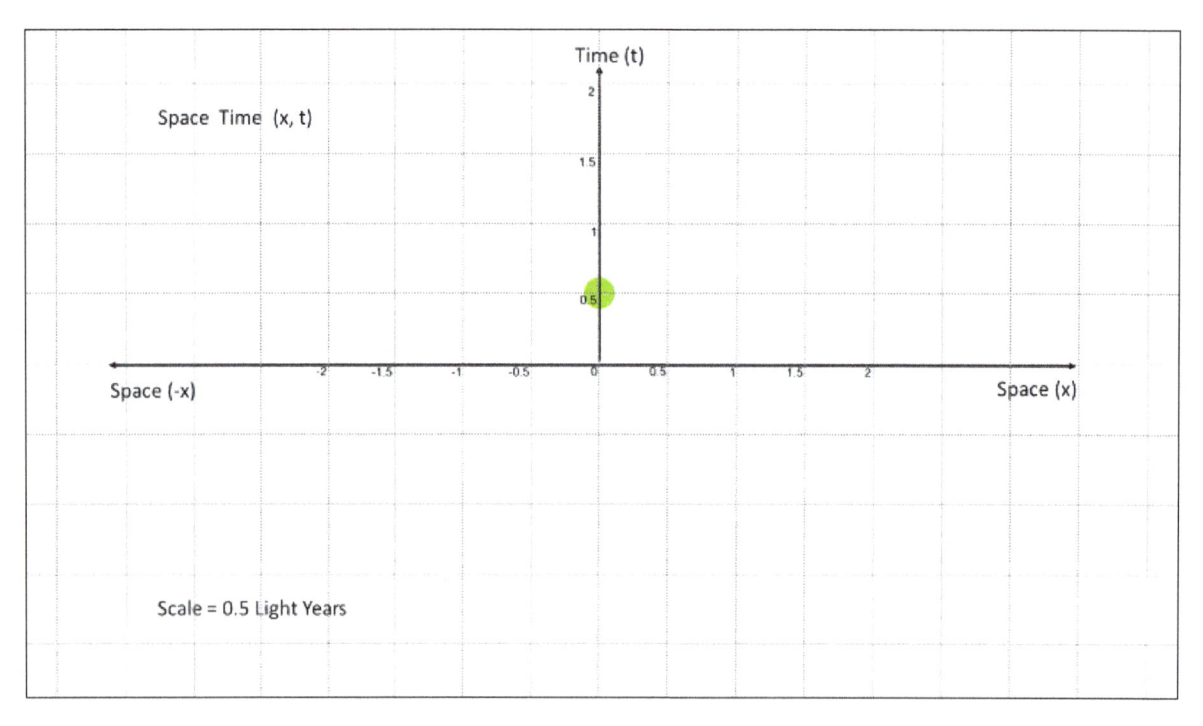

Figure 52.

If it sat on the launch pad for a further 6 months, it would be at the point x=0 and t=1

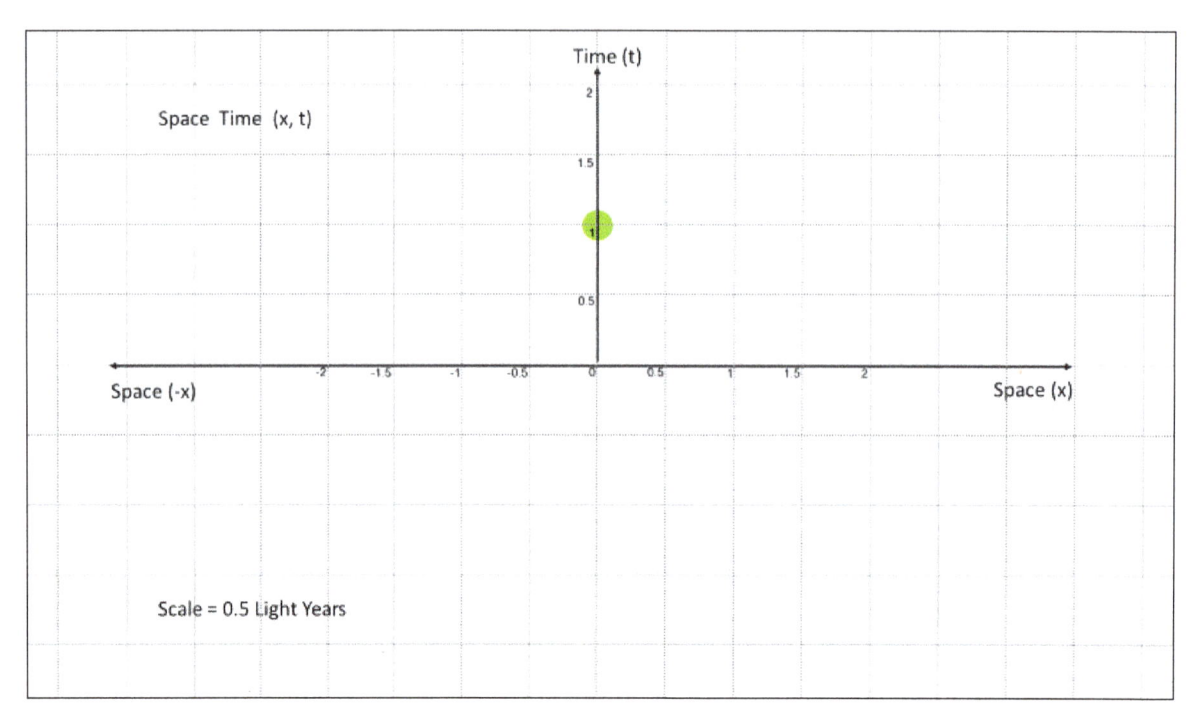

Figure 53.

Let's start over. Our spaceship is at x=0 and t=0. And we release a photon of light at x=-1, t=0. So, the photon – our light source – was released at a distance along the x-axis of 1 light-year away. We know that light will travel 1 light-year in 1 year. And so, a photon released at the point x=-1 and t=0, would after 1 year be at the point x=0 and t=1. It would have travelled 1 light-year in the x direction and taken 1 year in the t direction to do that. The orange marker is the photon released at x=-1 and t=0.

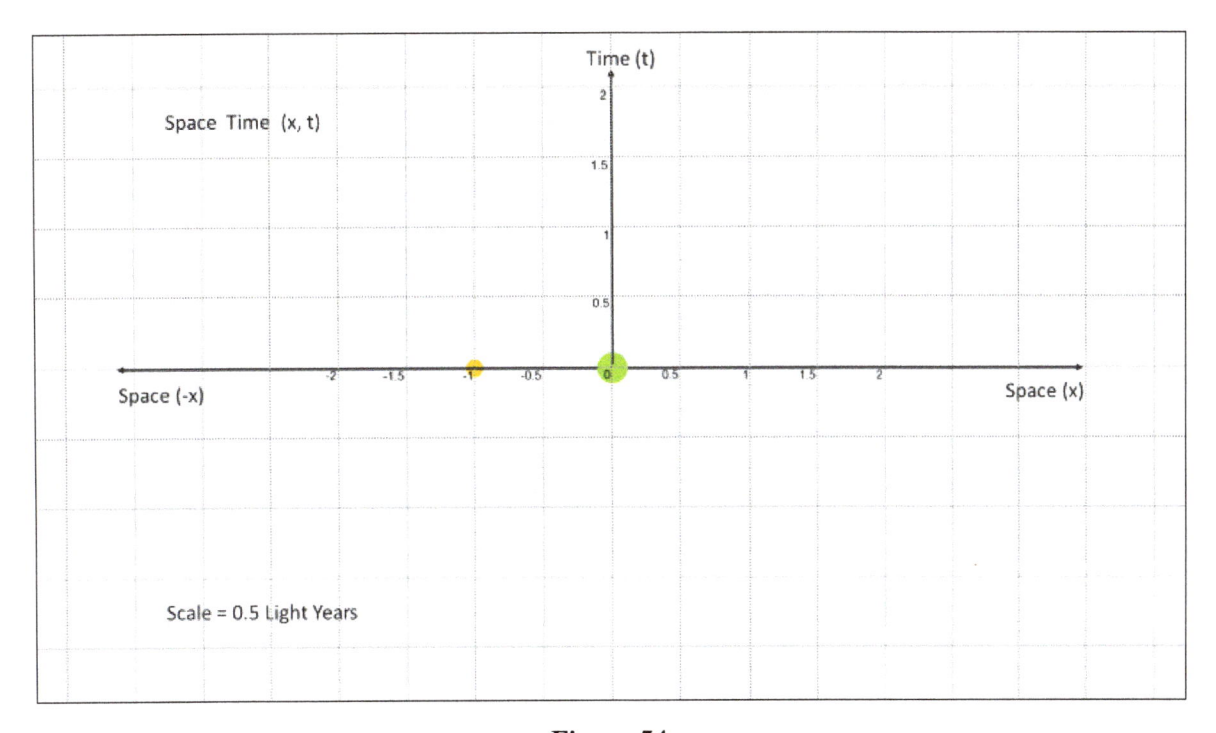

Figure 54.

Next, we see where the photon and spaceship are after half a year. The grey markers show where the photon and spaceship have been at earlier times. We might imagine the grey markers tracing the path taken through spacetime by the photon and the spaceship to get to where they are now after half a year has passed. Later, we will see the path light traces is somewhat different from what we see here.

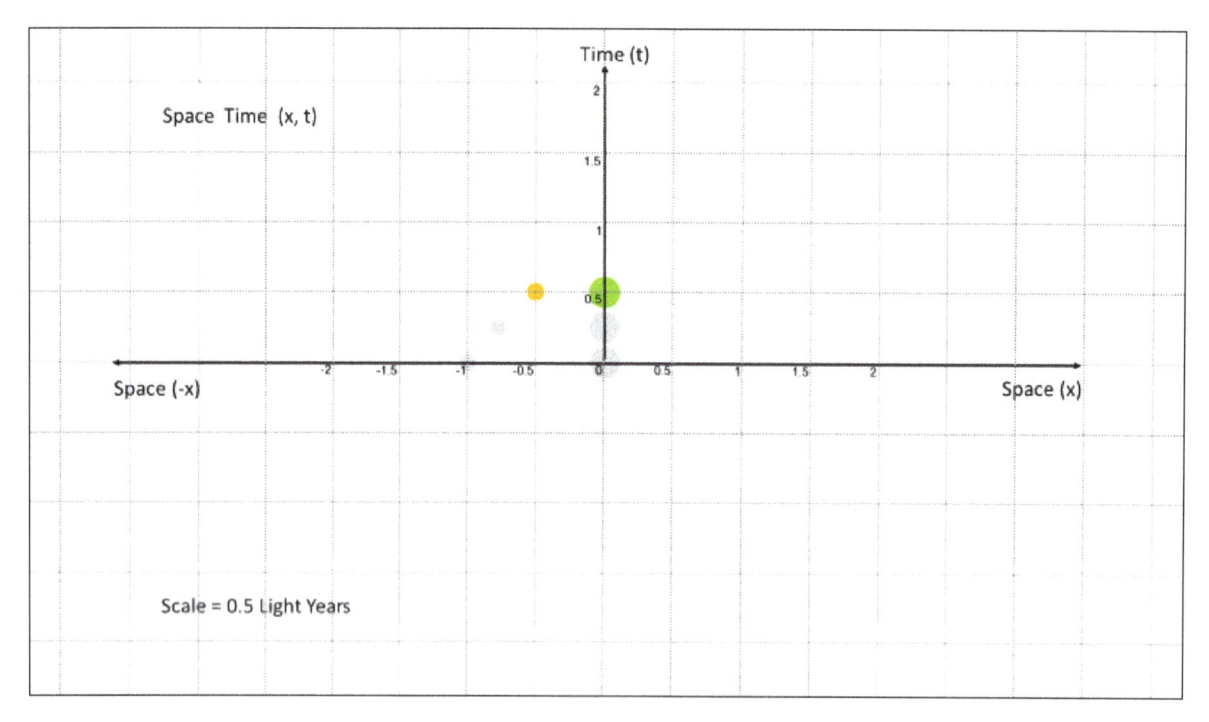

Figure 55.

What is light?

Now, we see where the photon and spaceship are after 1 year. It is at this point that an observer on the spaceship would see the light. Not before, and importantly – not after. Again, the grey markers trace the paths taken by the spaceship and photon through spacetime. For the photon, the grey markers show where the photon has been. But as we shall see, the path traced by light is *different* from the straight line we could make by connecting its grey markers.

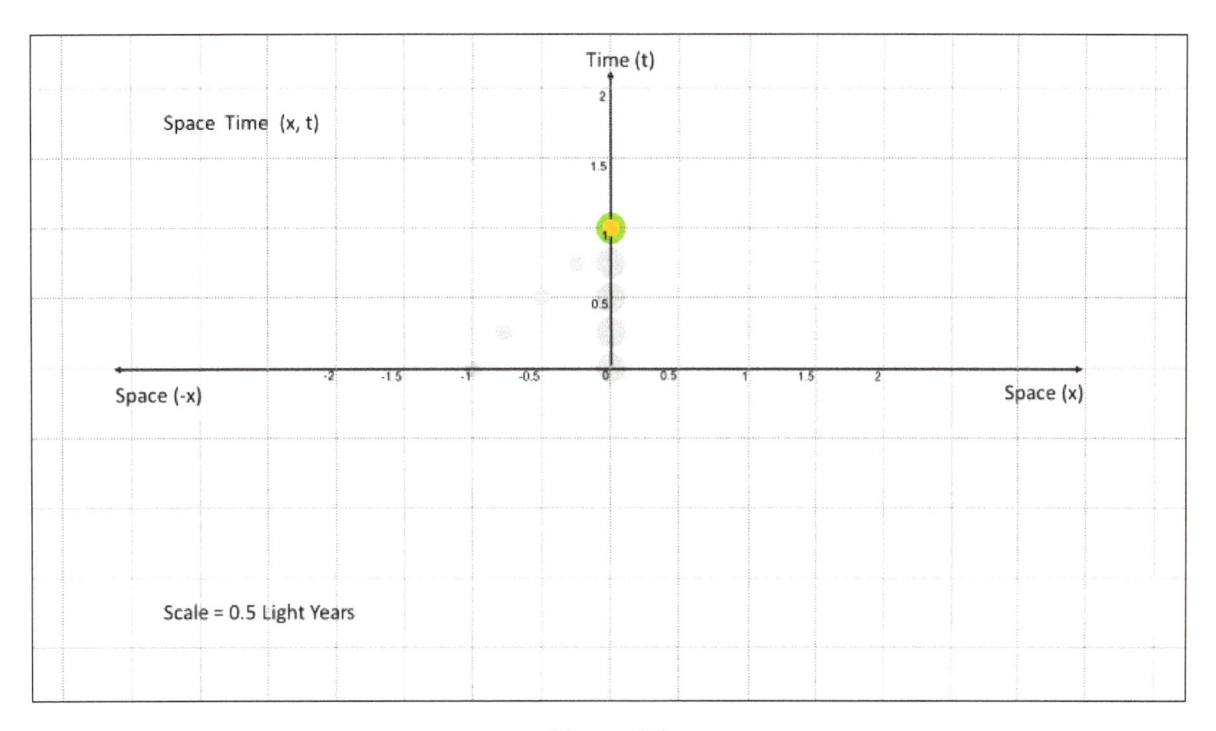

Figure 56.

When moving away from light

Let's start over. This time we have added a second spaceship using a red marker. It is flying away at half the speed of light in the x-direction. So, after 1 year it will have travelled half a light-year.

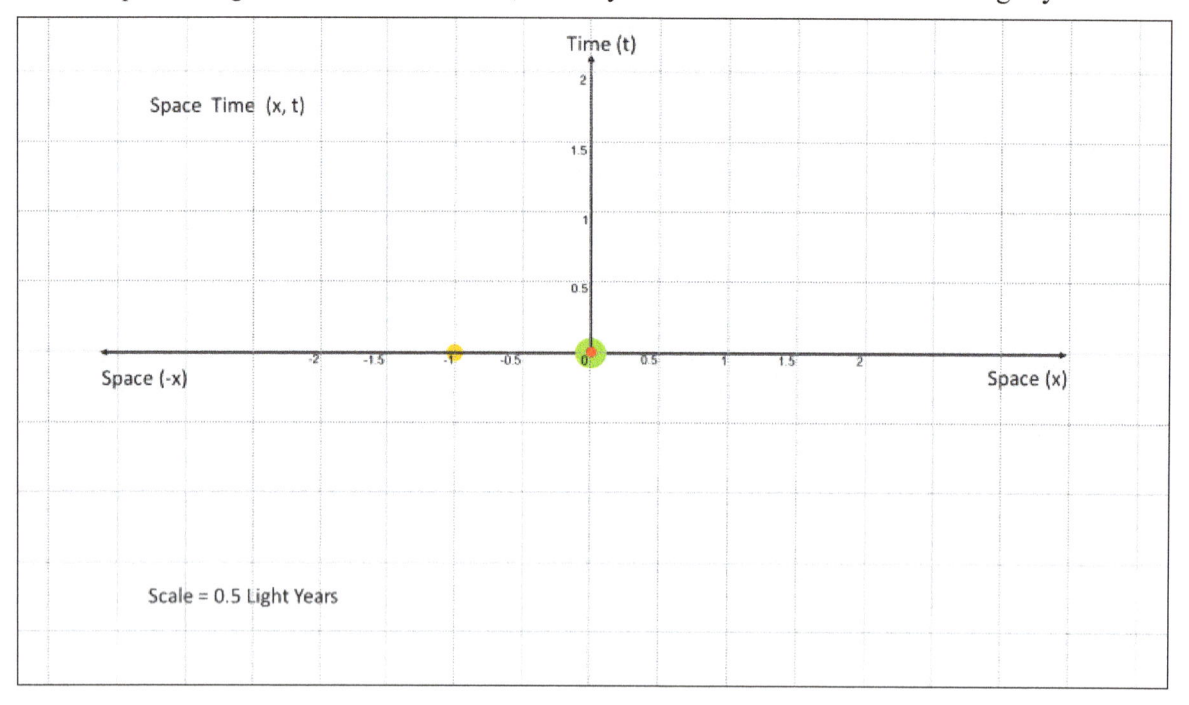

Figure 57.

After 1 year, our green spaceship that did not move is at x=0 and t=1. Our photon of light is at x=0 and t=1 and can be seen by our green spaceship. And our red spaceship is at x=0.5 and t=1 having travelled half a light-year in distance over the period of 1 year because it was traveling at half the speed of light. An important observation by our green spaceship is that it measures the speed of light as being 1 light-year per year or around 300,000 kilometres per second in accordance with the standard model. All is as it should be.

What is light?

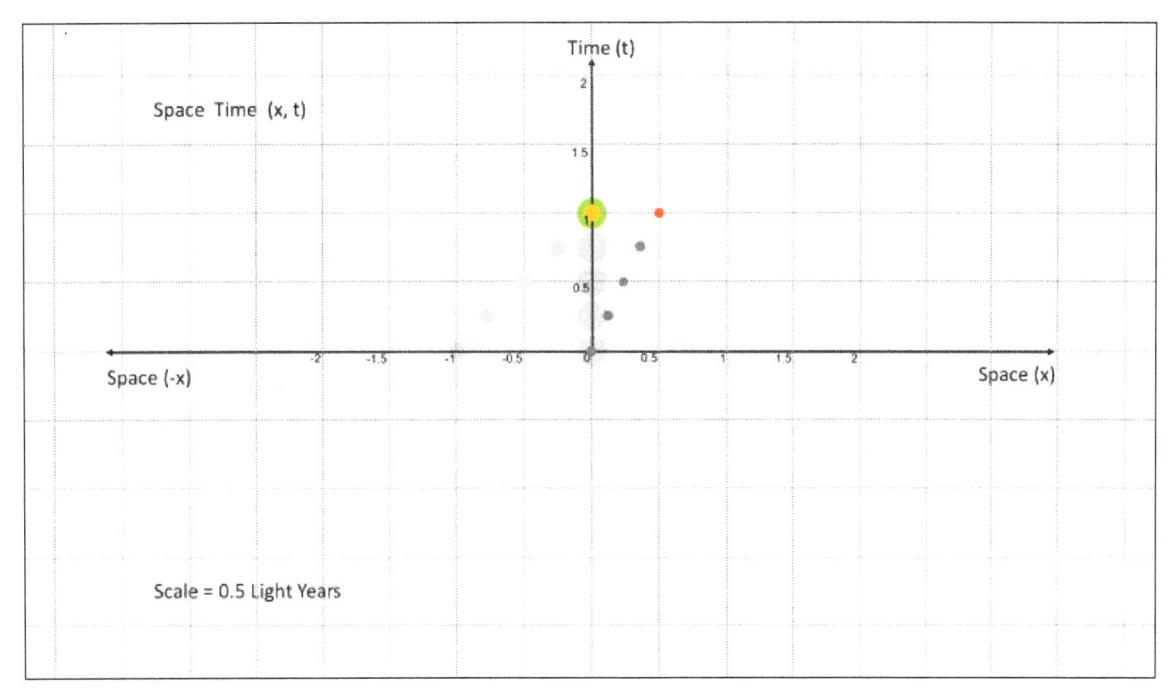

Figure 58.

After one and a half years our green spaceship moves on through time. Our red spaceship continues to fly at half the speed of light and is at x=0.75 and t=1.5. Our photon (had it not been observed by our green spaceship but passed by and continued) is at x=0.5 and t=1.5

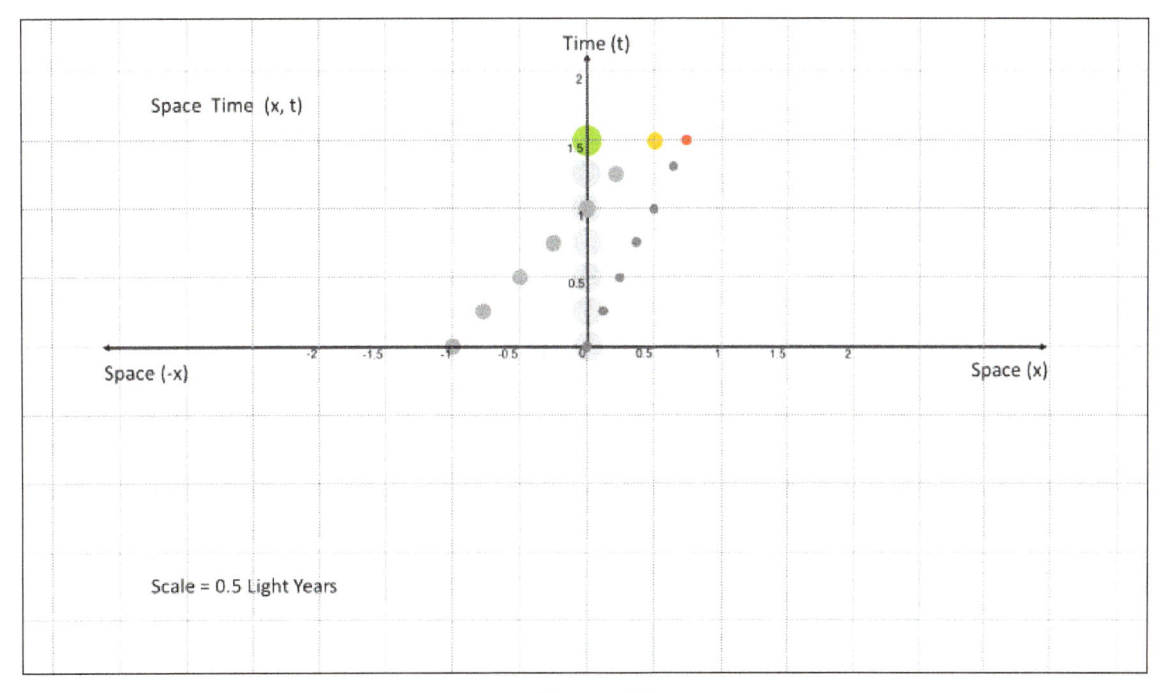

Figure 59.

115

Finally, after flying for 2 years, observers on our red spaceship see the photon that left the position on the x-axis at x=-1 and t=0 two years earlier. The observers on the red spaceship measure the approach speed of the photon. The photon is travelling in the x direction at light-speed. Had the red spaceship been *stationary*, it would have measured the speed of the photon as 2 light-years in 2 years, so 1 light-year per year, or in other words, light-speed.

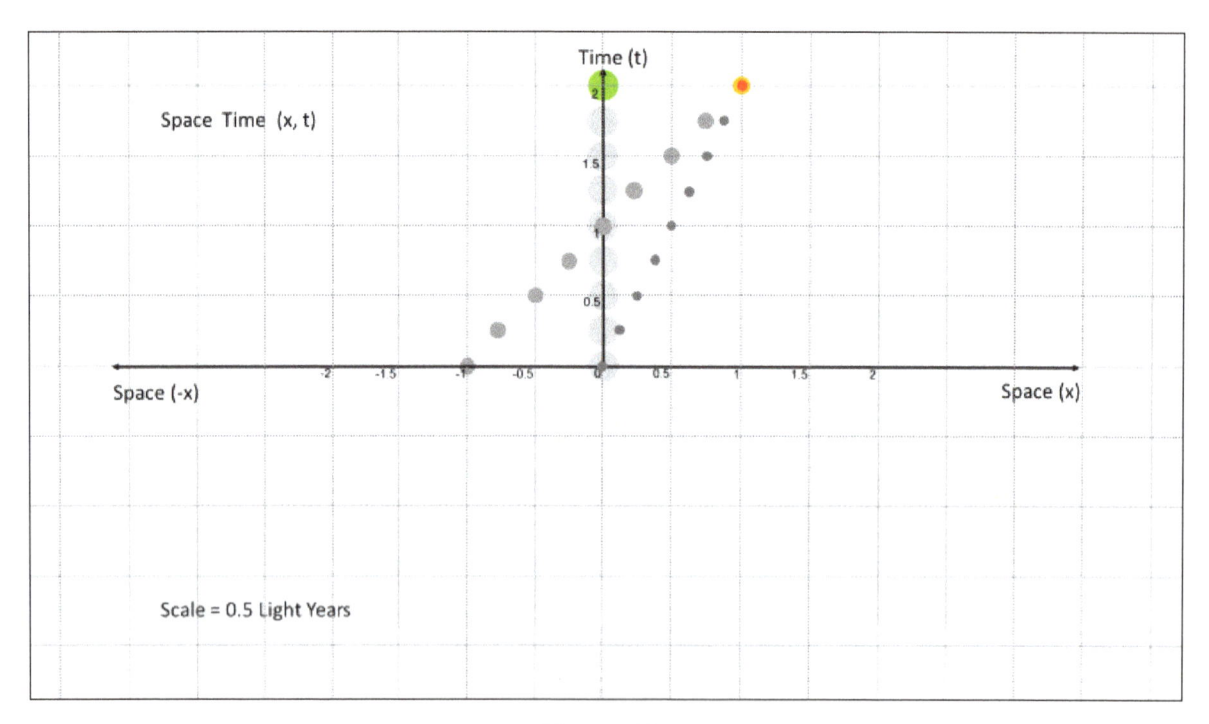

Figure 60.

But the red spaceship is flying *away* from the photon in the x direction at half light-speed. It therefore measures the *approach* speed of the photon at light-speed minus half light-speed (the speed the spaceship is flying away from the approaching photon). The value measured then, for the approach speed of the photon, is half light-speed.

But this is not the value demanded by the standard model. Regardless of the motion of the spaceship, the approach speed of light must always be light-speed. And it is! We know this to be true, because in the real world, when we do measure it, that's the value we always get: light-speed. And we get that value if we are standing still or moving. And we get that value regardless of how fast we are moving. And we get that value when we are travelling away from the photon or toward it.

So, we must conclude that our graph is wrong – somehow.

To understand where it's *wrong,* we first need to look at where it's *right*.

In the real world, our green spaceship travels through spacetime as it does in our graph. And it encounters the photon in the graph at the same time and place as it does in the real world, at x=0 and t=1. So, for our green stationary spaceship, moving through time, our photon presents at locations on our graph that match the time and place locations in the real world. In respect of the photon and our green spaceship then, our graph appears to be *right*!

How we move through spacetime

The only other item in our graph is the red spaceship. So we begin with the premise that the path it traces through spacetime as it flies at half the speed of light through our graph, is not how it flies through spacetime in the real world. We already know this from real-world measurements. In the real world, the faster we travel the slower time progresses. The effect is that the red spaceship cannot arrive at x=0.5, t=1; nor at x=1, t=2. Even as its travelling at half light-speed, it simply can't take a whole year to travel half a light-year, because when travelling at speed (any speed!), time slows down. At least that's what the math tells us. And real-world measurements confirm it.

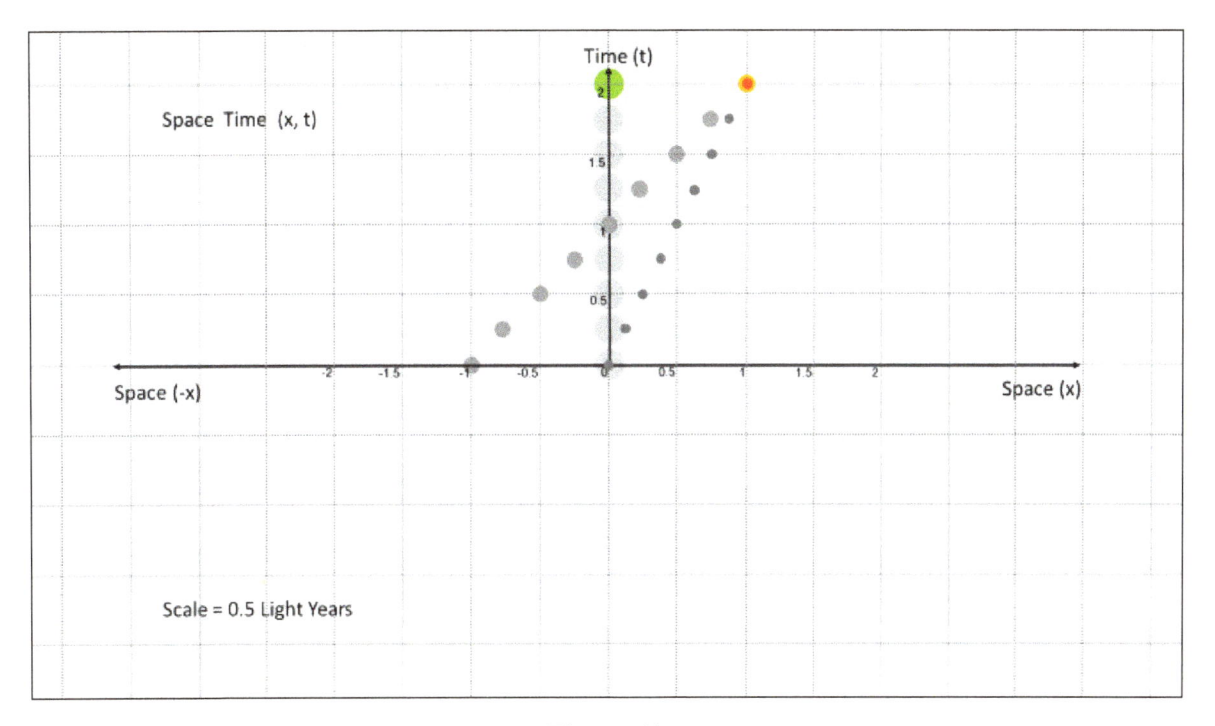

Figure 61.

Now, slowing down time for a travelling spaceship presents a real problem. It is easy to see that we cannot make time go slower in our graph for the red spaceship without also making time go slower for the green spaceship. Recall, that in our model, time is the speed that the x-axis moves up the time-axis. If we made time go slower for the green spaceship by lagging the progress of the x-axis up the time-axis, that would make time go slower for the photon too – and we would want that, otherwise they would not meet in the graph at the correct place and time, at x=0 and t=1. But the photon is supposed to travel at a 'constant' speed. Slowing time down for the photon but having it arrive at x=0, t=1 to intercept the stationary green spaceship would violate its requirement to travel at its constant speed. Also, the idea of making time go slower presents its own problems. That's because we have faithfully taken a two-dimensional slice of spacetime with all of x moving through time as it does in the real world. No point mass of any significance is present to lag the progress of the x-axis up the time-axis (the mass of the spaceships we can regard as negligible, surely?). Lagging the x-axis for the green spaceship just because a red spaceship is travelling at speed? What if there were a white spaceship travelling at half the speed of the red spaceship? How much ought we to lag the x-axis for the green spaceship then?

So, if time does not *slow down*, what's happening?

The faster we go, the faster we go!

When we think about the real-world measurements of time slowing down as we travel faster, we see that there is a different way of interpreting that. If we say that 'the faster we go the faster we go', that will require a little more explanation. So, I am going to spend some time explaining what I mean by that.

Say we want to reach a destination that is one quarter light-years away. If we travel at one quarter light-speed, we expect to reach our destination in a year. Now what if it didn't take a year to reach our destination? What if it took less than a year? If we measured how much less, we could call that value Q-minutes.

If we were to travel at half light-speed, we would expect to reach our destination in half a year instead of the year we would expect to take travelling at one quarter light-speed. But what if instead of it taking half a year, it took less than half a year? If we measured how much less, we could call that value H-minutes.

We notice that we save different amounts of time getting to our destination one quarter light-years away depending on how fast we travel. We save Q-minutes if we travel at one quarter light-speed, and we save H-minutes if we travel at half light-speed. We also note that H-minutes is greater than Q-minutes. It's more than twice Q-minutes. Travelling double the speed saves us more than double the time. So, the faster we go, the more minutes we save. The faster we go, the less time than the expected time it takes. If we take less time to do the same distance, we are in effect going *faster*. Hence, the faster we go, the *faster* we go.

It is important to note that these distances and speeds are from our perspective looking at the graph. The graph is our 'universal' frame of reference. We are observers able to see what is happening in the slice of spacetime represented by our graph.

Recall our red spaceship travelling at half light-speed? We can show on our graph that the position of the red spaceship, after travelling two years, would need to be lower down on the time axis for the same position on the x-axis. Why? Because we went *faster* and therefore took *less time* to travel the *same distance*. We can say, that after two years, our red spaceship is at x=1 and t<2, (**t** is less than **2**). We don't know exactly what the value along the t-axis is, only that it is less than t=2 but we don't really know by how much less.

OK, say, for the sake of making an additional point that it is at t=1.75

What is light?

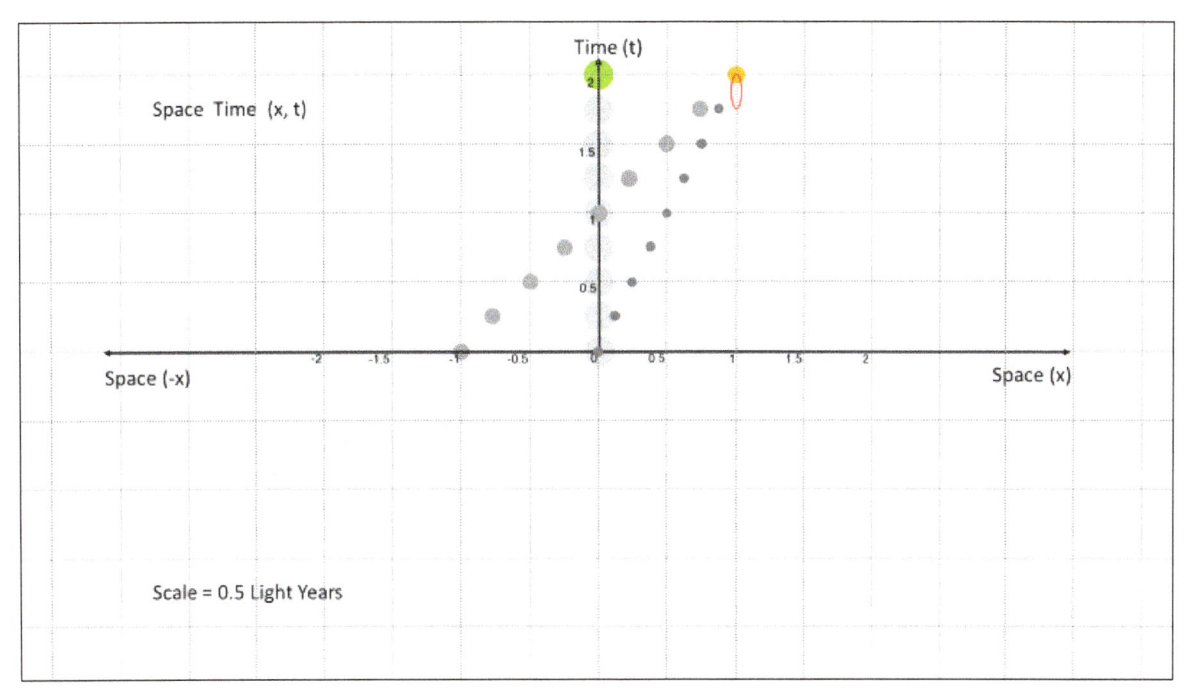

Figure 62.

Now this immediately presents a problem. We can see from our graph that if t=1.75 then our red spaceship will not encounter the photon at position x=1. The photon will need to travel longer than two years to encounter the red spaceship.

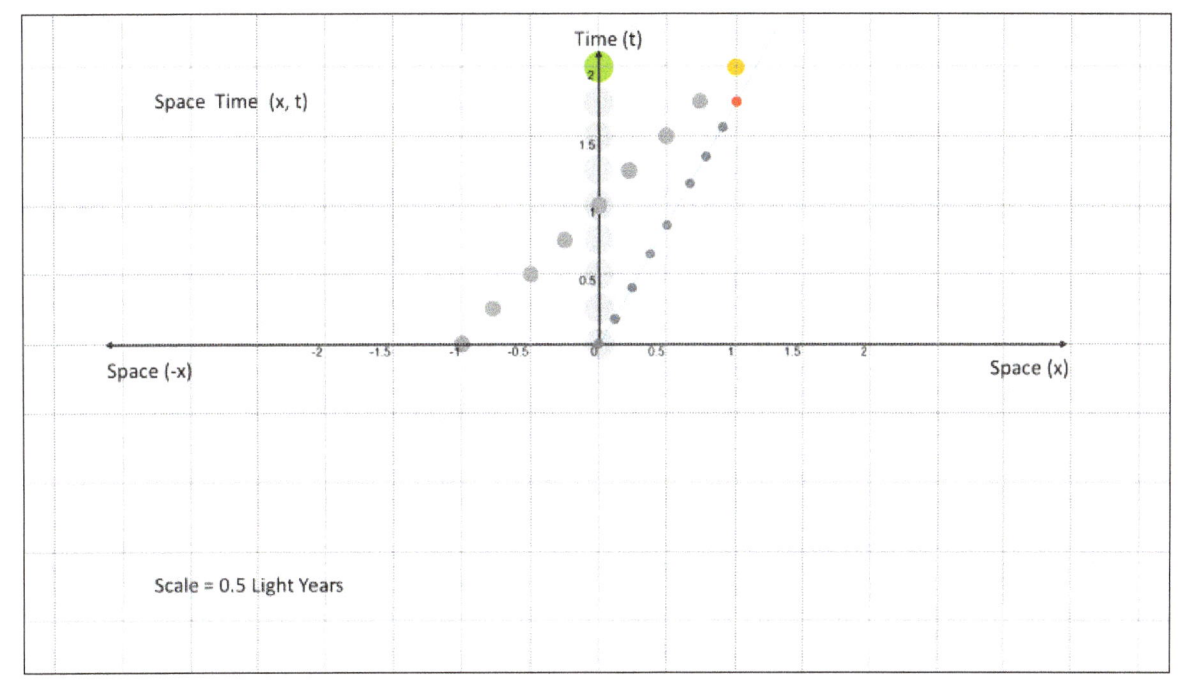

Figure 63.

119

We can estimate x=1.3 and t=2.4 as the spacetime location where the red spaceship would encounter the photon, if we said the faster speed of the spaceship found it at x=1 and t=1.75, and then photon and spaceship continued to travel until they met.

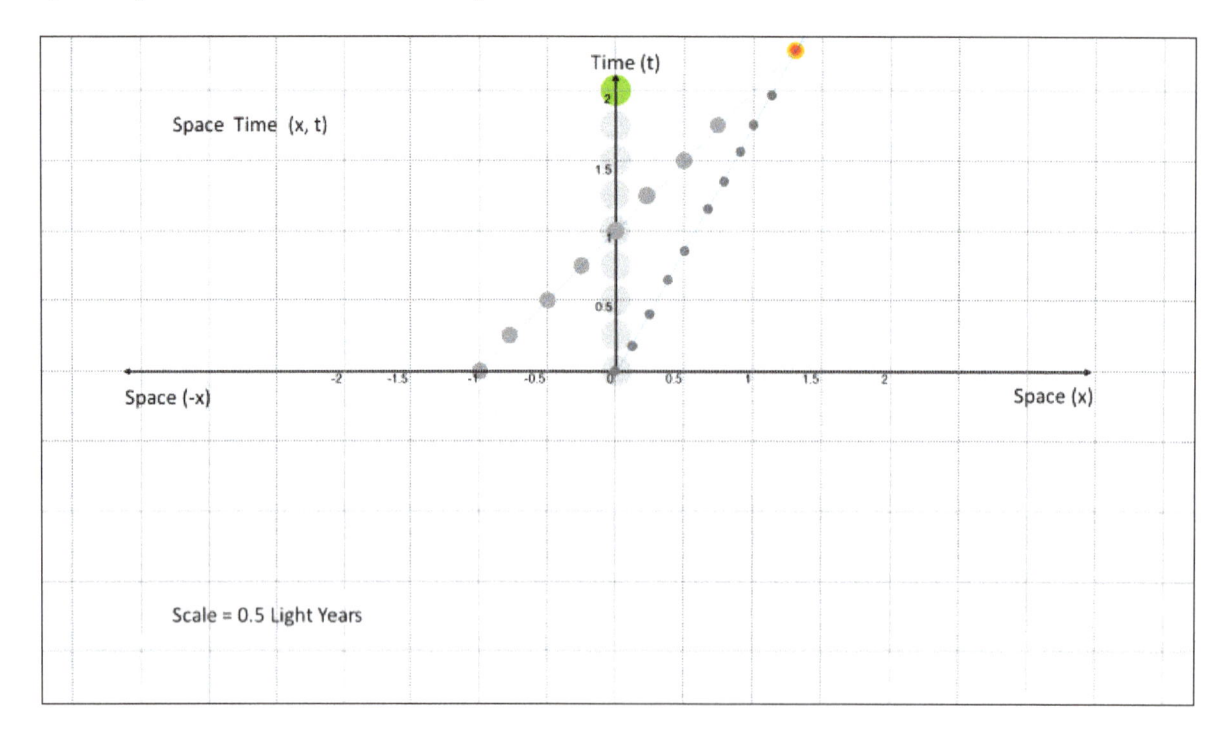

Figure 64.

But that does not solve the problem. Observers on the red spaceship will still not be able to calculate light-speed as the approach speed of the photon. The photon is still travelling at light-speed. The spaceship is still travelling away from the photon. Hence, light-speed minus the speed of the spaceship away from the photon will give the approach speed value. Hence, it will be a value less than the light-speed demanded by the standard model. Additional point made!

Getting there sooner: time-drops

If we take the position that indeed – the faster we go, the faster we go – then we must allow in the first instance that our red spaceship is lower down on the time axis than expected when it arrives at x=1. And if that's true, then all along its flight path it will be saving time. The further it travels, the more time is saved. This is shown in the graph below. Each blue arrow represents the *drop* in time taken to travel the distance at the red markers. The small grey markers pointed to by the arrows represent the spacetime locations due to the time drops.

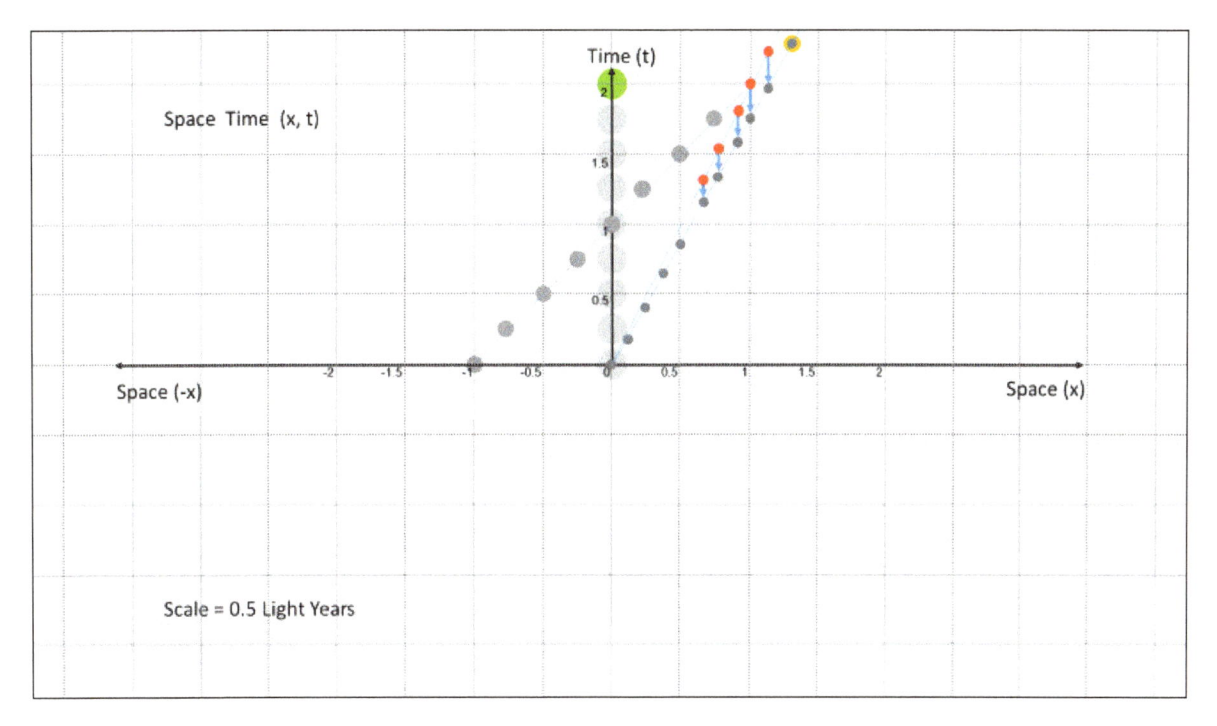

Figure 65.

Finding a photon's flight path

So, if we know how the spaceship travels through spacetime. If we know that as it speeds up, time *slows* down – and we do know this because we have measured it in the real world. And if we *interpret* that in the way we have, so as not to mess with the time-axis – indeed, so as not to mess with the time-dimension as it participates in the spacetime continuum. If we adhere to the slogan – the faster we go, the faster we go. Then we may need to consider that the flight path taken by the *photon* in our graph, as we have described it, isn't right. It's right for when the spaceship does not move, because it allows us to calculate the correct approach speed of light to be light-speed as demanded by the standard model. More importantly, light has that approach speed in the standard model because that's what we measured it to be in the real world. But in the real world, we also measured it to be light-speed when the spaceship was moving, regardless of whether it was moving toward the light or away from it.

Designing an experiment

The challenge, then, is to find a flight path for the photon so that its approach speed is always measured as light-speed. The way we do that is to consider the flight paths of not one, not two, but *several* spaceships, each travelling at a different speed away from the light source. We already know that less time is gained at lower speeds than at higher speeds. But regardless of the speed of the spaceship, light will eventually catch up with it. We can again use a light source that is exactly 1 light-year away from x=0 at x=-1. And release the photons at t=0 so that they commence their journeys when our spaceships are at x=0 and t=0, flying through that point at a constant speed, and at a different speed for each spaceship.

If we start with a spaceship travelling at a speed of zero, so that the spaceship is stationary, we have already seen where in spacetime the photon meets the spaceship. It meets it at x=0 and t=1.

We can then look at where the photon should meet a spaceship travelling at a faster speed. The criterion here, is that travelling at a faster speed, an observer on the spaceship should measure the approach speed of the photon and find it has a value of light-speed.

Photons: speeding them up

Now there is a real tendency to falter here. You see, for these calculations to come out right, we need our test photons to be travelling faster than light-speed. And we know that a photon always travels *at* light-speed. Indeed, we measured our photon travelling at light-speed at x=0 and t=1. Moreover, haven't we been told that nothing can travel faster than light-speed not even light itself? For those who have seen the equations in the standard model – it's thoroughly compelling. But the calculation for *approach speed* is the speed of the photon minus the speed of the spaceship (for when the spaceship is travelling *away* from the photon). That means the speed of our photon always needs to be light-speed *plus* the speed of our spaceship. When it is *that*, and we then do the approach speed calculation, i.e., the speed of the photon minus the speed of the spaceship, we always end up with a value for the approach speed being light-speed. And that's the value we want!

So, we need to increase the speed of *several* photons to discover the flight path a *single* photon would take. We send a *different* photon to each of our spaceships because each spaceship is travelling at a *different* speed. We need to speed up the photon for each spaceship by just the right amount, so that when observers on the spaceships measure the photon's approach speed, they get the value light-speed. We then note where along their flight paths our spaceships took their measurements and where in spacetime, they encountered their photon. We can then place a marker at those measuring points. One marker per spaceship flight path. We can then trace a line through those markers to plot the flight path that would need to be taken by a *single* photon, for all our spaceships to measure its approach speed to be light-speed. We can then see which path the photon took through spacetime, so that its approach speed will always be measured as light-speed, by our stationary and moving spaceships. That's a lot to take in. You might want to read through this paragraph again so that it's clear what we are going to do and why.

Measuring approach speed

We can plot the flight paths of spaceships travelling at zero, ¼, ½, and ¾ times the speed of light in blue. We can plot the paths of photons travelling at 1, 1 ¼, 1 ½, and 1 ¾ times light-speed in orange. We can then place red markers at the points where our spaceships encounter a photon travelling at a speed equal to the sum of light-speed and the speed of the spaceship, that is, a photon whose approach speed to it is light-speed.

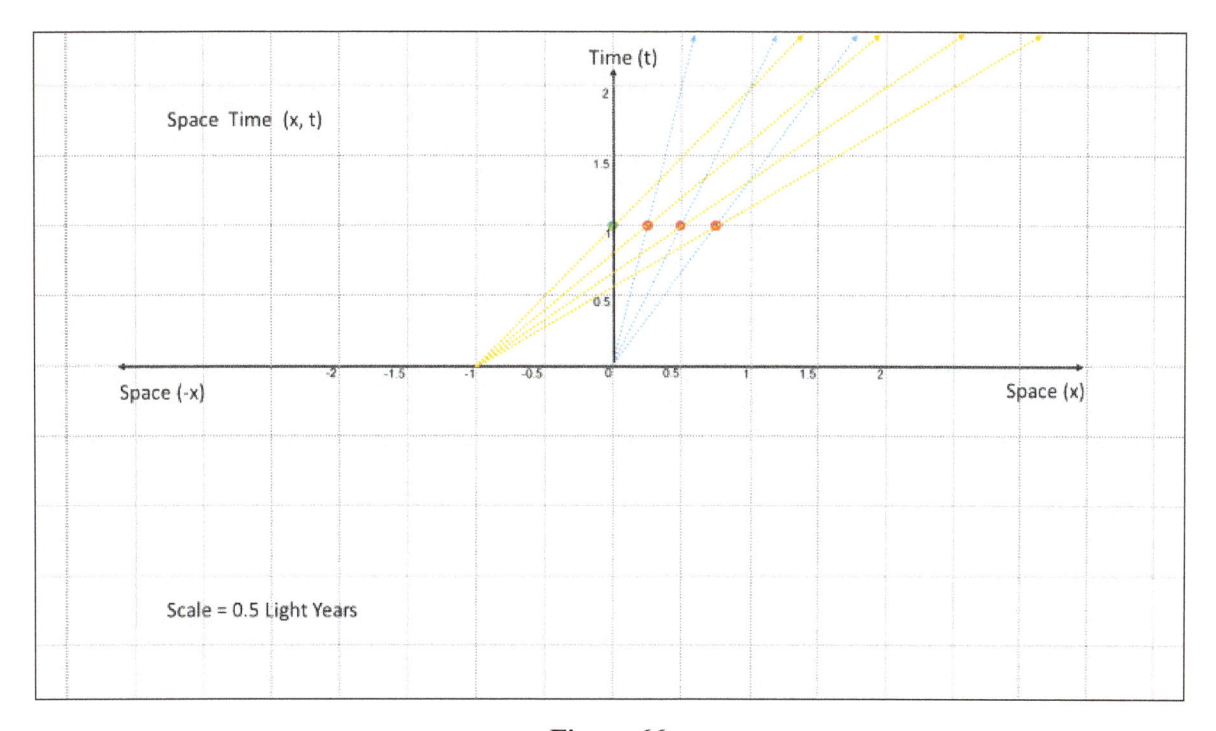

Figure 66.

What we see from the graph is that photons released 1 light-year away at x=-1, always intercept spaceships traveling away from them at t=1, regardless of the speed of the spaceship. Even the green spaceship that is stationary is intercepted at t=1.

This presents another problem. We release all four photons at t=0, that is, at the same time. All four spaceships were at or passing through x=0 and t=0 at the same time. Our goal was to derive the flight path of a single photon by analysing the intersection of four photons with four spaceships. We rigged the speeds of the photons to guarantee the approach speeds at the intersection points would always be light-speed.

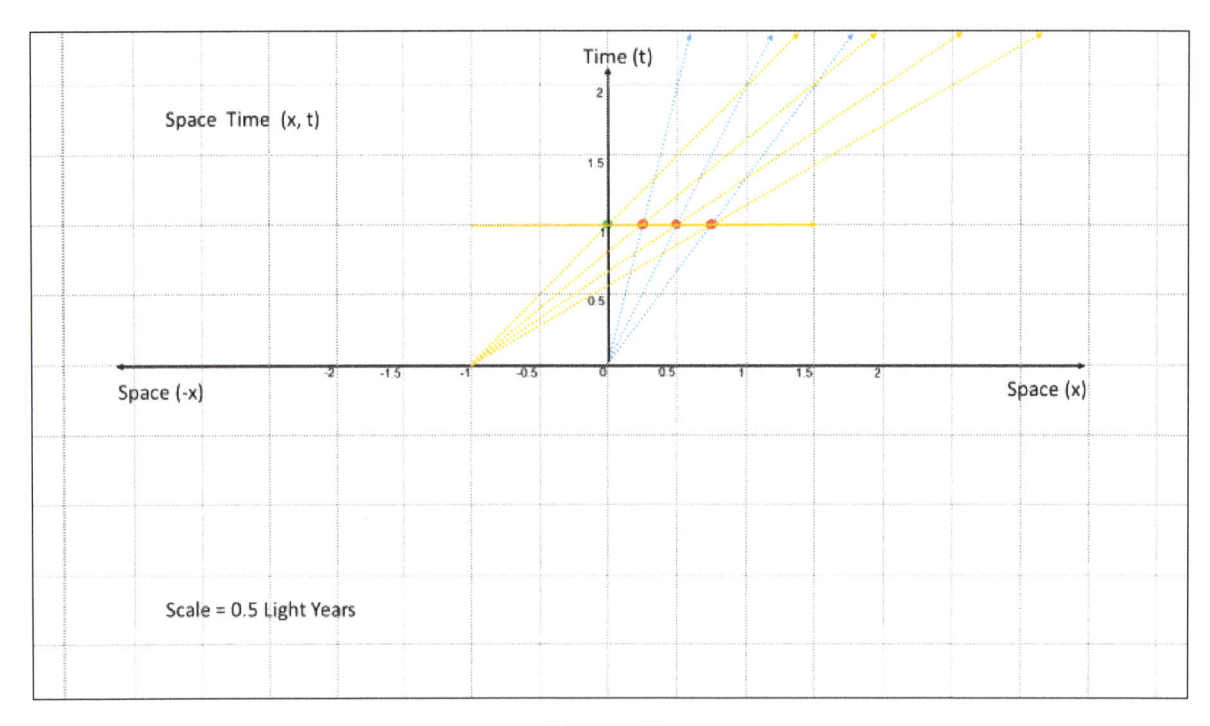

Figure 67.

But when we plot the flight path of a *single* photon (the solid orange line) that would fly through these intersection points, we find that it could never have been at x=-1, t=0. It was at x=-1, but the photon is not even moving through time – it is stuck at t=1. The photon flight path we are seeking is one that starts at x=-1 and t=0 but also passes through the intersection points we have labelled with our green and red markers. So, this experiment is either an epic failure or we did not set it up properly.

Incorporating time-drops

Recall the blue downward pointing arrows in a previous graph? They represented the drop in time taken to travel the distance along the x-axis at the red markers. We had not taken these time drops into consideration in our experiment.

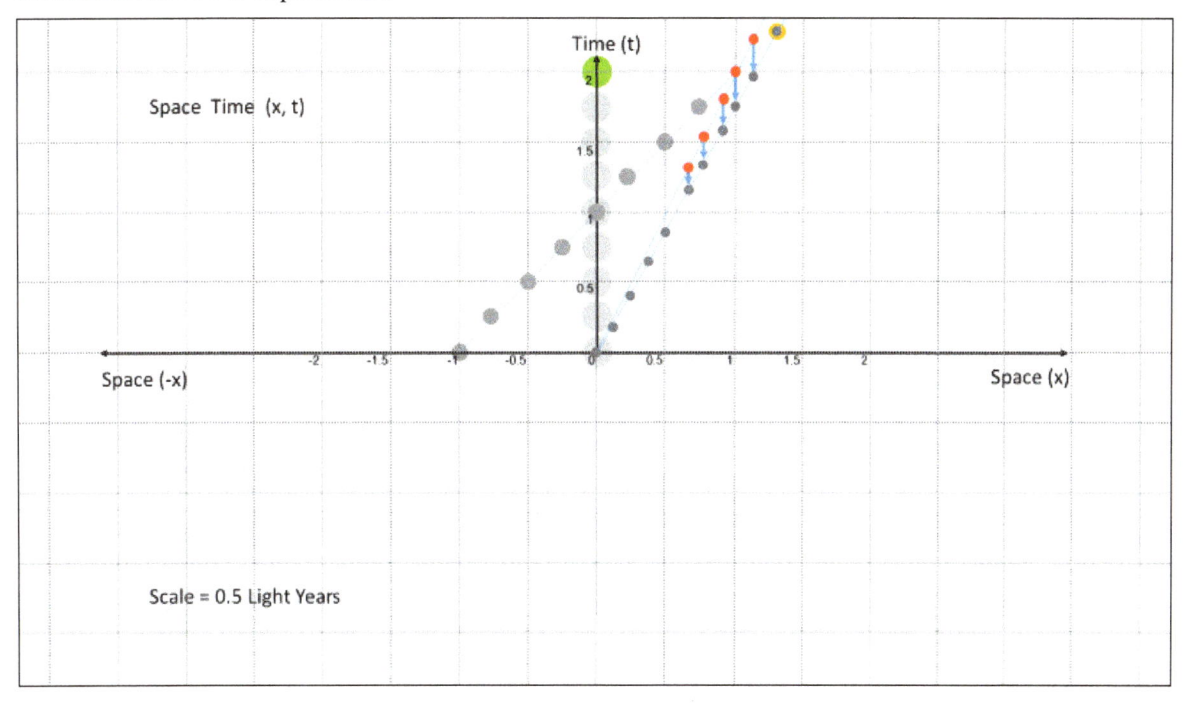

Figure 68.

If we applied the time drops to the flight paths of our four spaceships, what would their time-dropped flight paths look like? And remember – the faster we go, the faster we go!

Time-dropped trajectories

Here is the original chart with a few changes. We have removed most of the markers and the photon flight paths to reduce clutter. We are left with the original flight path of a spaceship travelling at half light-speed, passing through the point x=0.5 and t=1. We have the time-dropped flight path as a dotted line.

What is important to notice here is how much to the right the time-dropped flight path is at t=1. There are two ways to measure the time drop. One way is to measure the vertical distance between where the spaceship was on its original flight path and where it would be after the time-drop. The other way is to measure how much further the spaceship would have travelled if it had spent the same amount of travel time as the spaceship that did not experience the time-drop. In other words, when the time-dropped spaceship is at t=1, where would it be at x? The graph shows the extra distance travelled in the x direction as a red bar between the points of two black arrows. The red bars represent the distance gained. Distance is gained all along the flight path. It is the distance gained that allows us to reach our destination in less time.

If the faster we go, the faster we go, then we would expect the time drops to be greater and the red bars to be longer at faster speeds. At slower speeds, we would expect the time drops to be less and the red bars to be shorter.

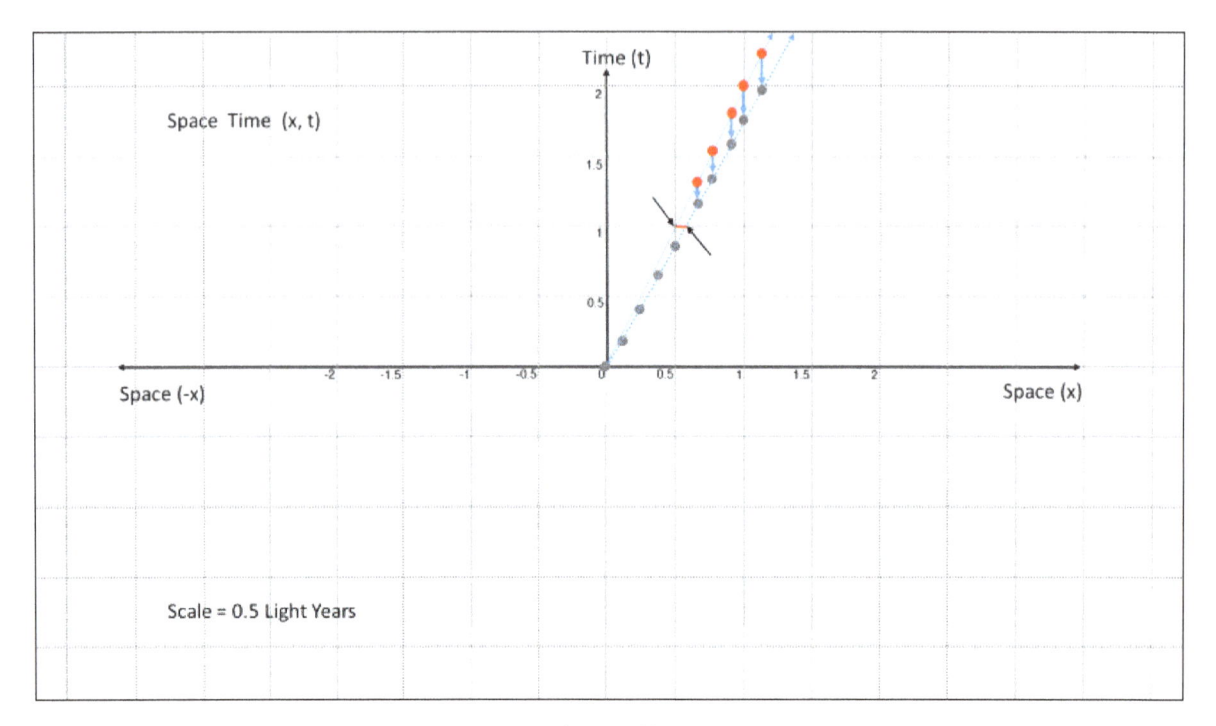

Figure 69.

We also know from the standard model that when travelling at a speed which is light-speed, time stops! Science tells us that we really can't travel at light-speed, but that *if we could*, time would stop. That may sound mysterious but it's what the science is currently telling us. We can show the effect of that in our graph by honouring that speed with a time-drop to zero. Doing *that* also gives us a clue about how much change there is in the size of the time drops as we go faster and faster from a standing start.

There is no time-drop in the trajectory of our *stationary* spaceship moving through spacetime. The time-drop for a spaceship if it could travel at light-speed would be 1 and see our spaceship at t=0. In between, the time drops get progressively larger as we increase our speed from zero to light-speed.

We also see that the angle made by the trajectory of a spaceship travelling at light-speed is 45 degrees with the x-axis as it passes through x=1 and t=1. That suggests the length of the red bars must get longer much more quickly as we progress through higher speeds than when progressing through lower speeds. Increasing speed does not gain a proportional increase in the time saved. It is as though when you accelerate from one speed to the next, additional acceleration is generated by the time-drop.

The next chart shows time-drops as the left sides of red triangles. Some blue arrows remain from the previous graph to help you see what I've done.

The extended distance travelled, had the spaceships continued travelling for a duration equivalent to the time-drop is shown here as the top sides of each triangle.

The hypotenuse of the triangle, its diagonal side, is the trajectory of the *time-dropped* spaceship.

Now, I did cheat a little bit. Because I have done this before. Rather than guess or estimate the size of the time-drops for the different speeds of the spaceships, I calculated them – exactly. I also exactly calculated the time-dropped flight path trajectories for our spaceships travelling at their different speeds.

The reason I did the calculations before drawing the graph is that I want you to see something important when you look at it. A graph made using calculated values makes that easier to see.

I could have estimated or guessed the time-drops, the time-dropped flight paths, and the extended travel distances. But we would have needed to create a lot more flight paths, each at a different speed, to reveal the flight path of that single photon.

That's what I did originally to develop this graph. I drew many flight paths. Each one for a spaceship travelling faster than the last.

Because I knew that slower speeds had smaller time-drops than faster speeds, no time-drop that I guessed could be smaller than for a slower flight path, nor larger than for a faster flight path.

Also, it turned out that the sizes of the time-drops are equivalent to the sizes of time dilations calculated under special relativity for spaceships travelling at velocity. And that's what we would expect. So, they are the calculations I used to get the exact values of the time-drops. That brings us back to two different perspectives. One perspective is that time slows down the faster we travel. Another is that the faster we go, the faster we go – we use less time because we arrive at our destination sooner, having travelled faster than we thought we did. It really doesn't matter which perspective you take. Either perspective will draw us the graph and will reveal the information we seek with our experiment.

As I increased the number of time-drops in the graph, the trajectory of that single photon began to emerge.

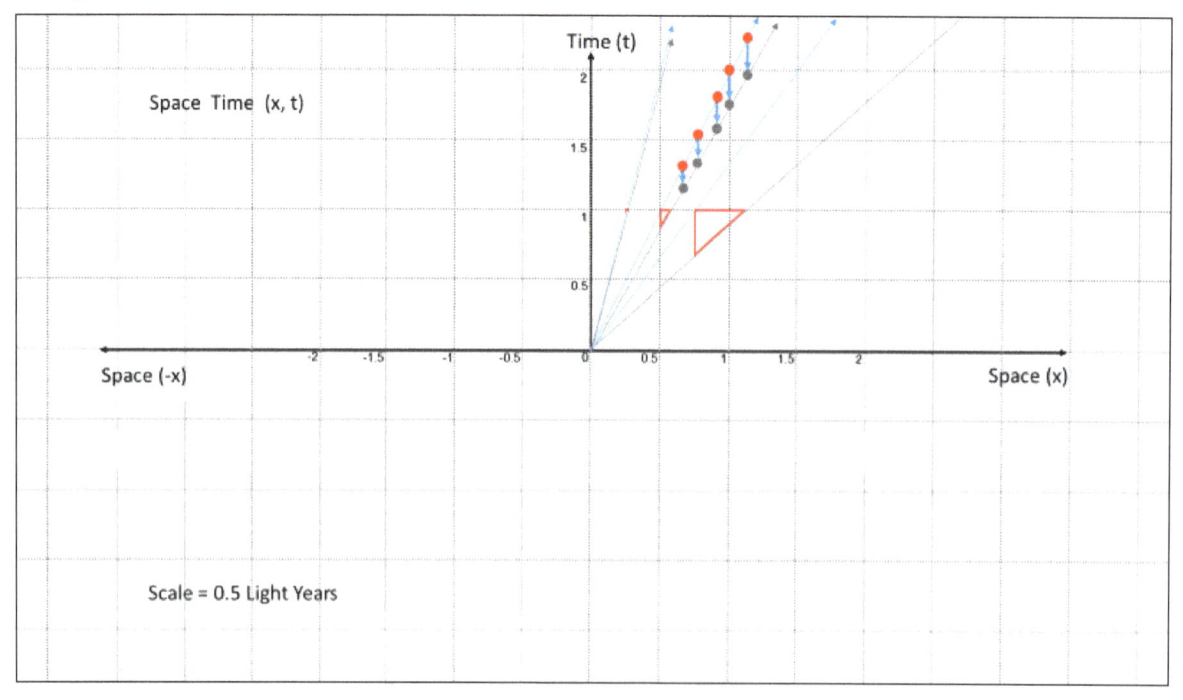

Figure 70.

Now that we have an accurate graph of the time-drops, I'd like to place a marker at the bottom point of each triangle, and then remove the triangles.

I have also included the time-drop marker for the stationary spaceship and the time-drop marker for the spaceship that travelled at light-speed, theoretically. What we notice about these markers is that they do not make a straight line when we join them together.

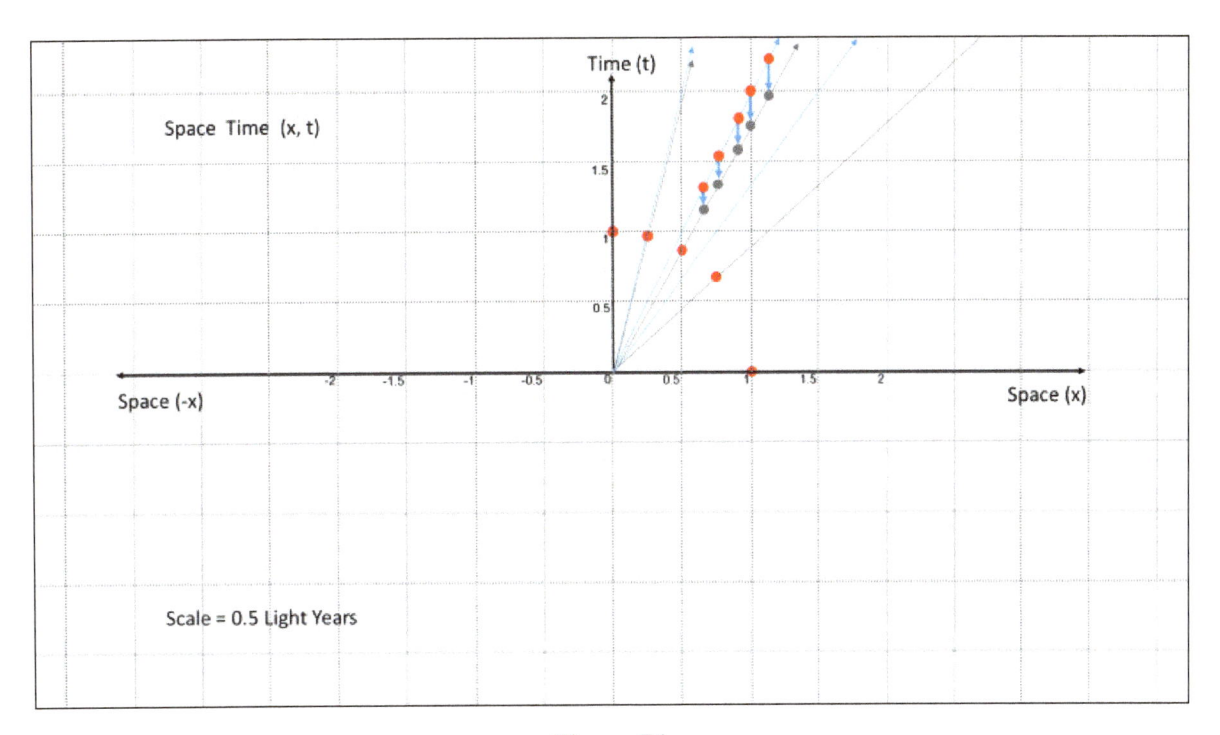

Figure 71.

They appear to subscribe a curve, if not the top right quarter of a circle. We can draw a circle with its centre at x=0 and t=0 to see if that's true.

A photon's curved flight path

Wow!

Our time-drops all landed on the circumference of a circle. The origin of the circle is at x=0 and t=0. And the radius of the circle is 1. In the time direction the radius is 1 year. In the x direction the radius is 1 light-year. What have we discovered?

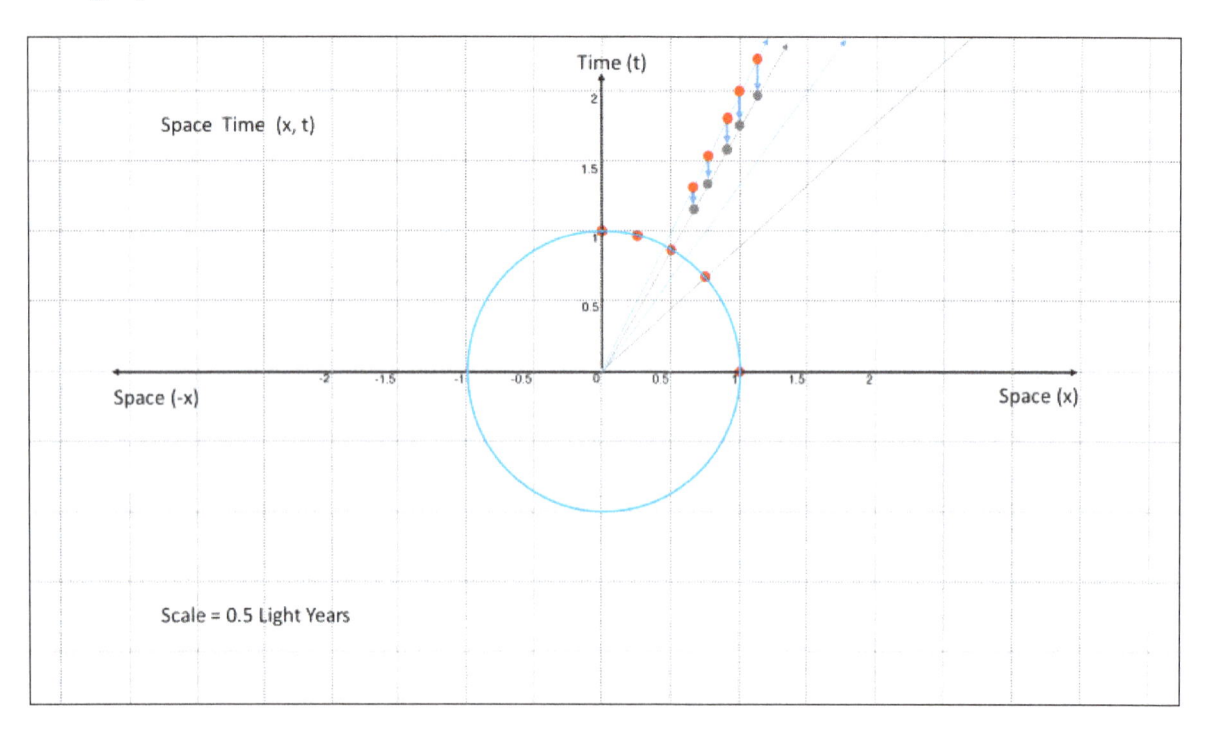

Figure 72.

Our aim was to discover the flight path of a single photon, which if encountered by spaceships travelling at different speeds away from it, would measure the approach speed of the photon as a value equal to light-speed. So, what does our graph tell us?

It's a bit difficult to imagine but we might be able to say that the photon started at x=-1 and t=0. At least if somehow this circle is its flight path, it does go through that point. An observer on our *stationary* spaceship would know the photon was released from a location 1 light-year away. The photon is observed at t=1 and so, 1 year has passed. The measured approach speed of the photon would have a value equal to light-speed. It has travelled 1 light-year in 1 year.

But what about observations made on our moving spaceships?

The fastest one, travelling at ¾ light-speed has an initial blue flight path that intersects x=0.75 and t=1. But due to the time-drop, (or time dilation value), resulting from its speed, its grey flight path is the time dropped one it is actually travelling on. On that flight path, it encounters the photon at x=0.75 but at an earlier time.

The crew on the spaceship know they have been travelling at 0.75 light-speed and have at that exact moment travelled 0.75 light-years away from x=0 and t=0. So, they conclude they have been travelling for 1 year. But, when they look at their clock, they see that less than 1 year has passed.

Time dilation?

There's a real conundrum here.

On one hand, physics tells us that a body moving at constant velocity will continue at that velocity until acted on by a force. So, this spaceship was travelling at a constant velocity of 0.75 light-speed when it passed through x=0, and t=0. We are therefore motivated to believe that the moment it passed through the x=0.75 light-year point, it was still travelling at 0.75 light-speed. But having travelled 0.75 light-years from x=0, the clock on the spaceship ought to read t=1. One year ought to have passed. But less than a year had passed. The spaceship had travelled the 0.75 light-years in less than 1 year. Therefore, somehow, it's speed when passing the x=0.75 light-years point must be greater than 0.75 light-speed.

And of course, that does not bode well for the rule in physics that demands constant speed remain constant.

Interestingly, the crew on the *stationary* spaceship note that the faster spaceship had indeed been travelling for one full year at the time it passed through the x=0.75 light-year point. But it too noted that the clock on the faster spaceship had slowed. So from the perspective of the stationary observer, the spaceship had travelled at 0.75 light-speed even as the clock on the spaceship had slowed.

And the crew on the faster spaceship also conclude they had travelled for one full year at 0.75 light-speed to reach their destination and fly through that point – even as their clock was showing that a year had yet to elapse.

This conundrum will be addressed by TRADD later. For now, let's go with the idea that the crew believed they were travelling at 0.75 light-speed. Because only if we do *that*, will the calculations come out right – even for the TRADD model.

Now, the crew of the faster spaceship, knowing where the photon came from, and that it started at x=-1 and t=0 (the same time that they flew through x=0 and t=0), they know that the photon, too, has been travelling for 1 year. Because the photon was encountered at x=0.75 they calculate that it had travelled 1.75 light-years in that time. That's the distance between x=-1 and x=0.75 where they are now.

They conclude that the photon has been travelling at 1.75 light-years per year. They subtract the speed that their spaceship is traveling away from it to get the approach speed and find it has a value of light-speed.

Each of the other spaceships do their calculations and come up with the similar results. The photon's approach speed to their spaceship is light-speed.

These *are* the results the standard model says we must get – because in the real world, when we measure the approach speed of a photon – of light – it is always light-speed. So, we are encouraged to believe that the flight path of the photon is curved and follows the circumference of the circle.

The odd result with respect to time here is that the clock on the stationary spaceship, as well as the clocks on the travelling spaceships, show that different amounts of time had passed at the precise moment their respective destinations were reached. Later, when we investigate what time is, this will all make sense. But for now, we can see from the graph that 1 year has passed for the stationary spaceship, but less than a year has passed – according to our graph – for the travelling spaceships on their time-dropped trajectories. Given that we calculate speed as distance divided by time, we do indeed see that the travelled spaceships had sped up – but only (perhaps) because their clocks had slowed down. Does that mean we need to rethink how we calculate speed and velocity? Or do we need to rethink what speed and velocity are? We do indeed do this later on, so stay tuned.

Fascinating stuff, isn't it?

Special relativity tells us that from the perspective of the stationary spaceship, the clocks on the travelled spaceships have slowed down – and our graph correctly shows this. But, as we will later see, from the perspective of any travelling spaceship, it is the clock on the stationary spaceship that has slowed. How can I say that? Well, if the travelling spaceship regarded itself as stationary, then the stationary spaceship, from the point of view of the travelling spaceship, would be moving. It does not matter which spaceship is travelling, or even if they are both travelling. The rate their separating distance increases defines the velocity of either spaceship away from the other – their relative velocity. This is an underlying principle of special relativity: time dilation from the perspective of an observer is a measurement dependent on the *relative* motion of the body being observed. Special relativity is aptly named because the message here is that measurements are always *relative*. But because our experiments have confirmed that it was the travelled clock that slowed, not the stationary clock – our universe is still hiding a reality that we have yet to come to grips with. We calibrate global positioning satellites' (GPS) clocks to account for their slowing down due to the speed they travel at – we speed them up to compensate. (We also slow them down to compensate for time speeding up because they are further out in the Earth's gravitational field.) We speed them up and slow them down to arrive at a net speed adjustment to account for the effects of first special relativity, and then general relativity.

When moving toward light

Almost there! That's all fine for spaceships moving away from the photon or standing still. But what about spaceships moving toward the photon? They too need to measure the approach speed of the photon to be light-speed.

Designing an experiment

In this experiment, we must reverse the direction our spaceships fly. This time they are flying toward the photon. The way we measure approach speed when travelling toward a photon is to add the speed of the photon to the speed of the spaceship. We want the result to have the value equal to light-speed.

Photons: slowing them down

But for that to happen, we need to slow our test photons down. We need the speed of our spaceships plus the speed of their test photons to equal light-speed. So, the spaceship travelling at ¼ light-speed needs to encounter a photon travelling at ¾ light-speed. The spaceship travelling at ½ light-speed needs to encounter a photon travelling at ½ light-speed. The spaceship travelling at ¾ light-speed needs to encounter a photon travelling at ¼ light-speed. And the stationary spaceship needs to encounter a photon travelling at light-speed. And if we *could* have a spaceship travelling at light-speed, it would need to encounter a photon standing still – not moving at all.

Measuring approach speed

Each spaceship drops a red marker when it encounters a photon having a speed, which when added to the speed of the spaceship equals light-speed. That's how we can find out where along the flight paths of our spaceships the test photons were encountered.

We find that all spaceships encounter their test photons at t=1.

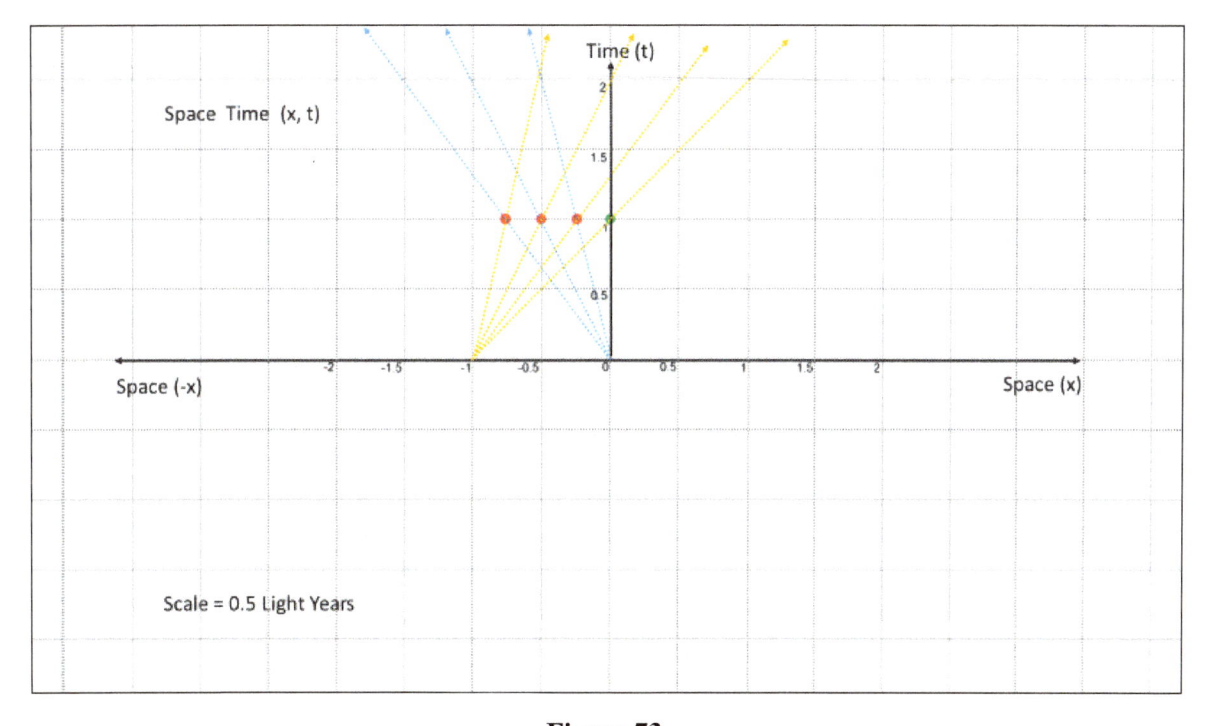

Figure 73.

Incorporating time-drops

Now, we have seen this result before when travelling away from the photon. It's because we used the flight paths from the flight plan. We have yet to consider that the faster you go the faster you go. We want to see where our test photons are encountered when our spaceships travel along their time-*dropped* flight paths.

Time-dropped trajectories

When the spaceships travel along their time-dropped flight paths the test photons are encountered along a curve; not along the straight line at t=1.

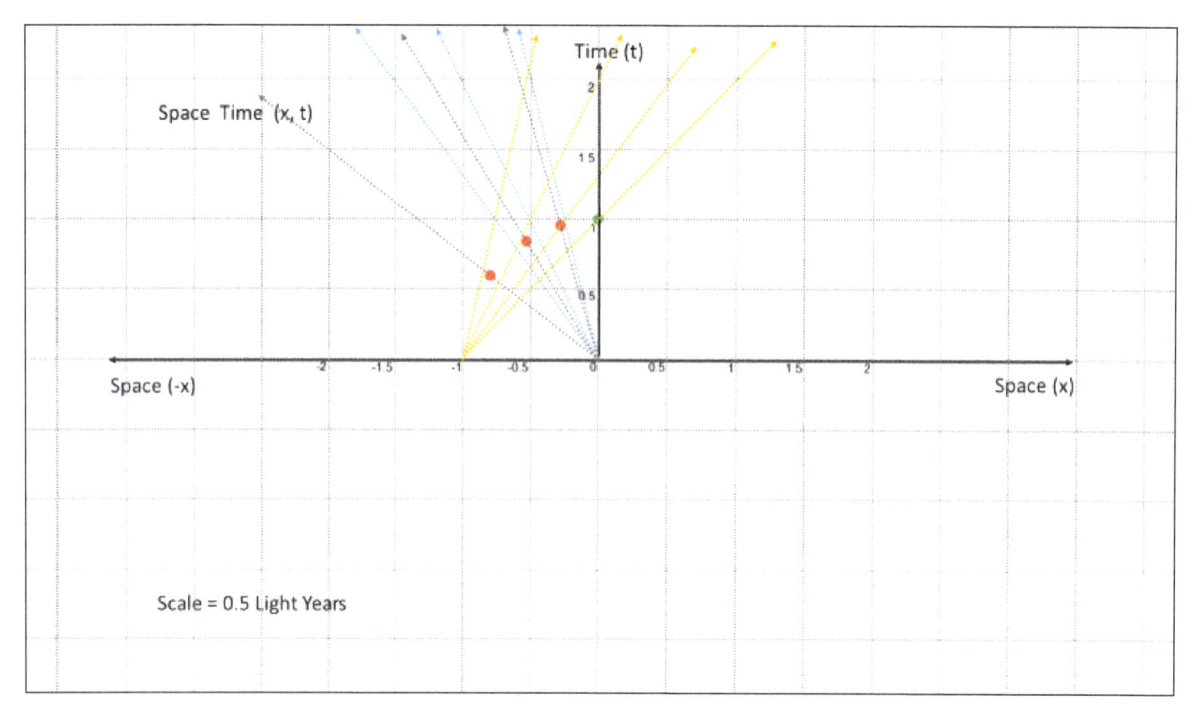

Figure 74.

We can see that performing time-drops when travelling toward our test photons works in the same way as when we were travelling away from the photons. The only change is the speed of the test photons. We had to slow them down.

A photon's curved flight path

In the graph below, I have added the marker from the spaceship theoretically travelling at light-speed. And I've included the circle that defines the curve along which the test photons were encountered. Again, we conclude that for a *single* photon to encounter at light-speed any of our spaceships travelling toward it, regardless of *their* speed, that photon would need to have a flight path following the curve along the blue circle's circumference.

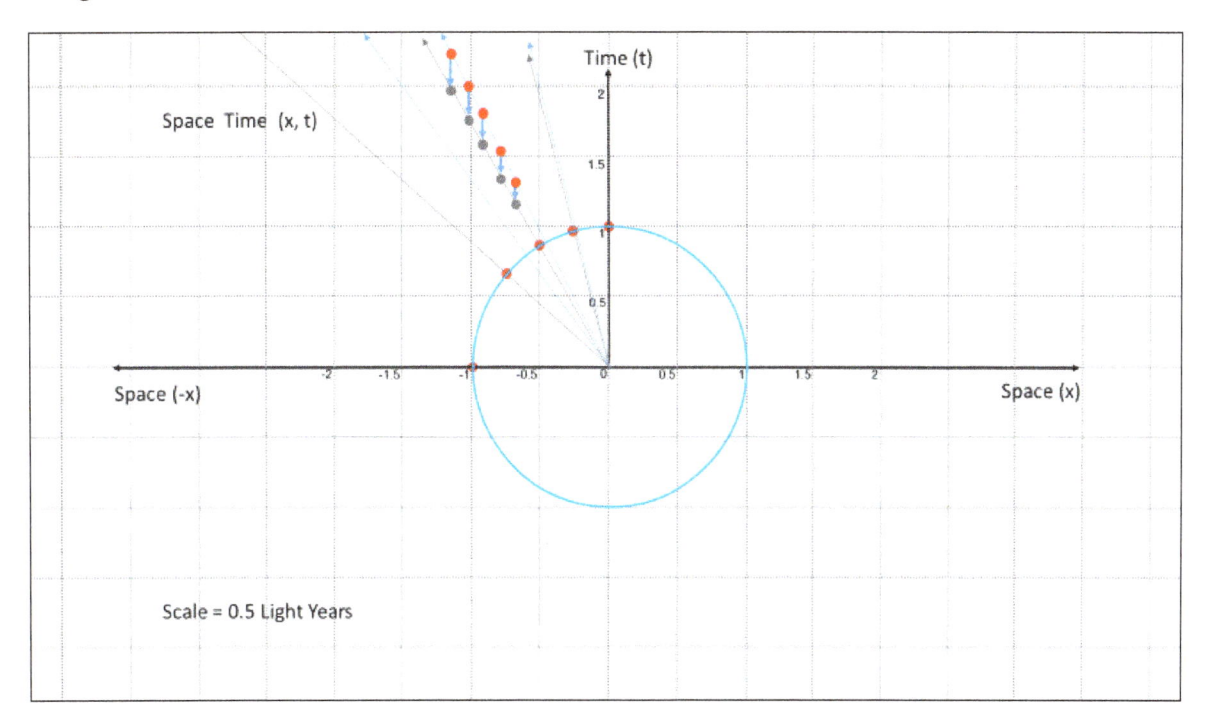

Figure 75.

Toward, away from, or waiting for a photon

We have produced graphs of spaceships moving away from and toward the photon. In each graph, we determined that the flight path of the photon must follow the curve made by a blue circle. Let's now put the two graphs together. We can see that the blue circle that defines the curve along which the photon travels – its flight path – is the same for when our spaceships are travelling toward the photon as for when our spaceships are travelling away from the photon.

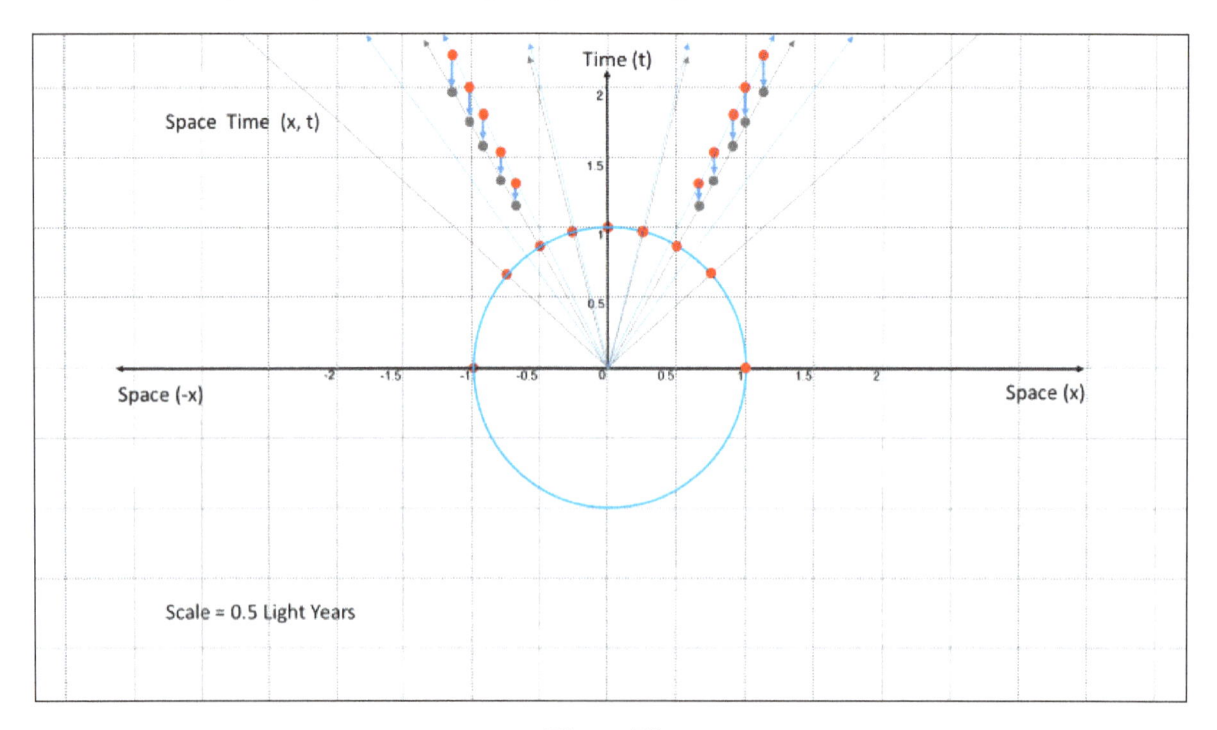

Figure 76.

More questions!

But there's still a problem! To show you what it is, we first need to unclutter our graph by removing our spaceship flight paths and the time-drop graphics.

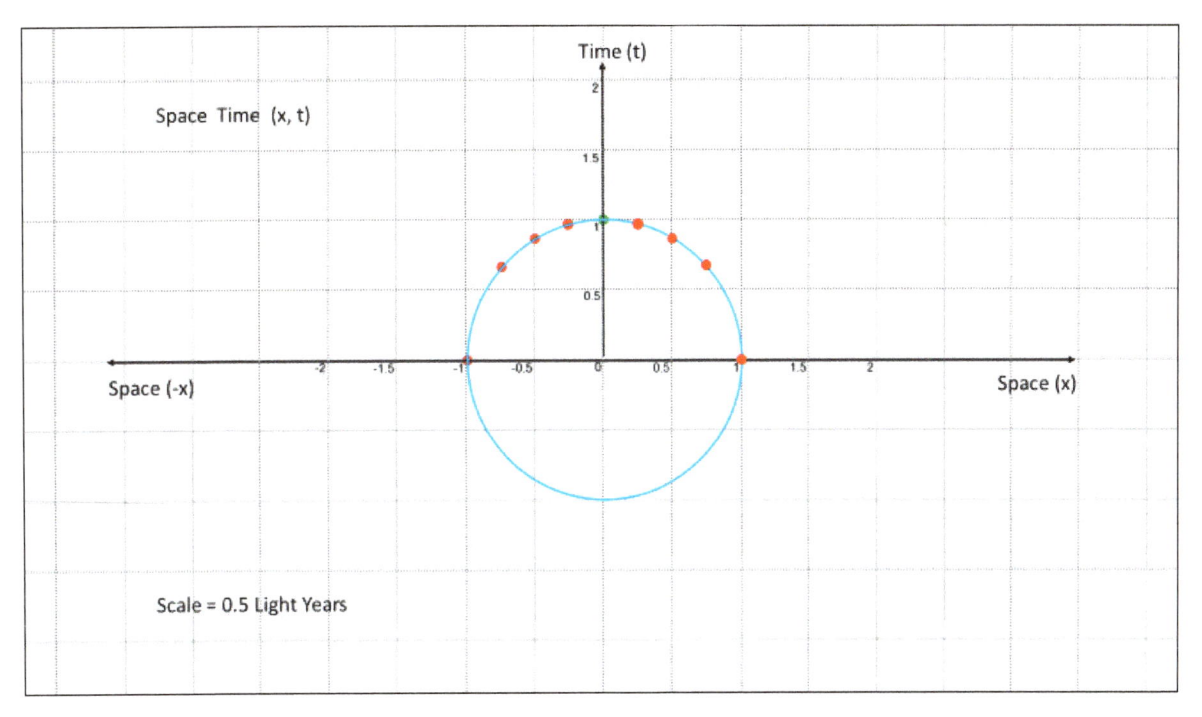

Figure 77.

Waiting for a photon

If we say that the photon travels along a flight path that follows the outline of the blue circle as shown in the graph, then that leaves us with an interesting question. How does our stationary spaceship see the photon after sitting on the launch pad for, say, half a year? Well, it doesn't. Because the photon has not arrived yet. It takes light a year to reach x=0, t=1 and so the only time it will be seen at x=0 is at t=1. The chart below shows where the photon would be after half a year as the smaller circle. It reaches up to t=0.5 when the stationary spaceship is at t=0.5 but has not arrived at x=0, t=0.5.

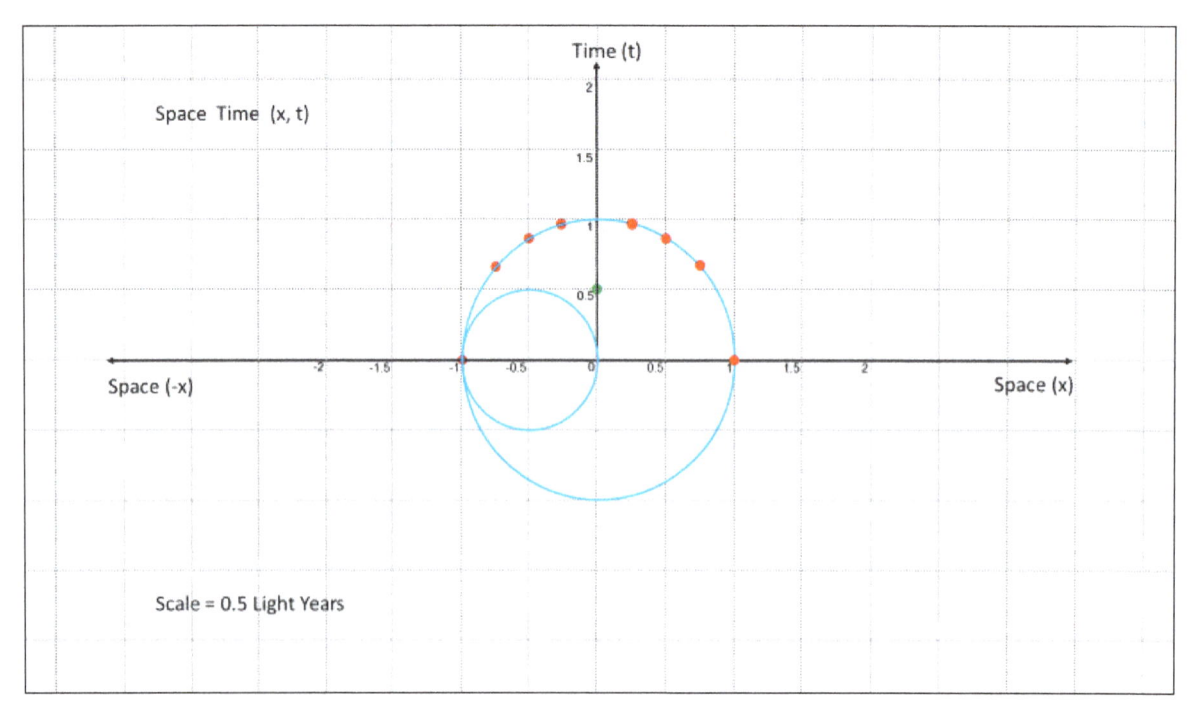

Figure 78.

Photons from less than a light-year away

What about light that is less than 1 light-year away? Do the rules governing when we would see the photon still apply when the photon is emitted, at say, 0.75 (purple) and 0.5 (green) light-years away? We can see that they do. Our spaceships encounter the purple photon after the photon has been travelling for ¾ of a year. And the green photon after it has travelled for ½ a year.

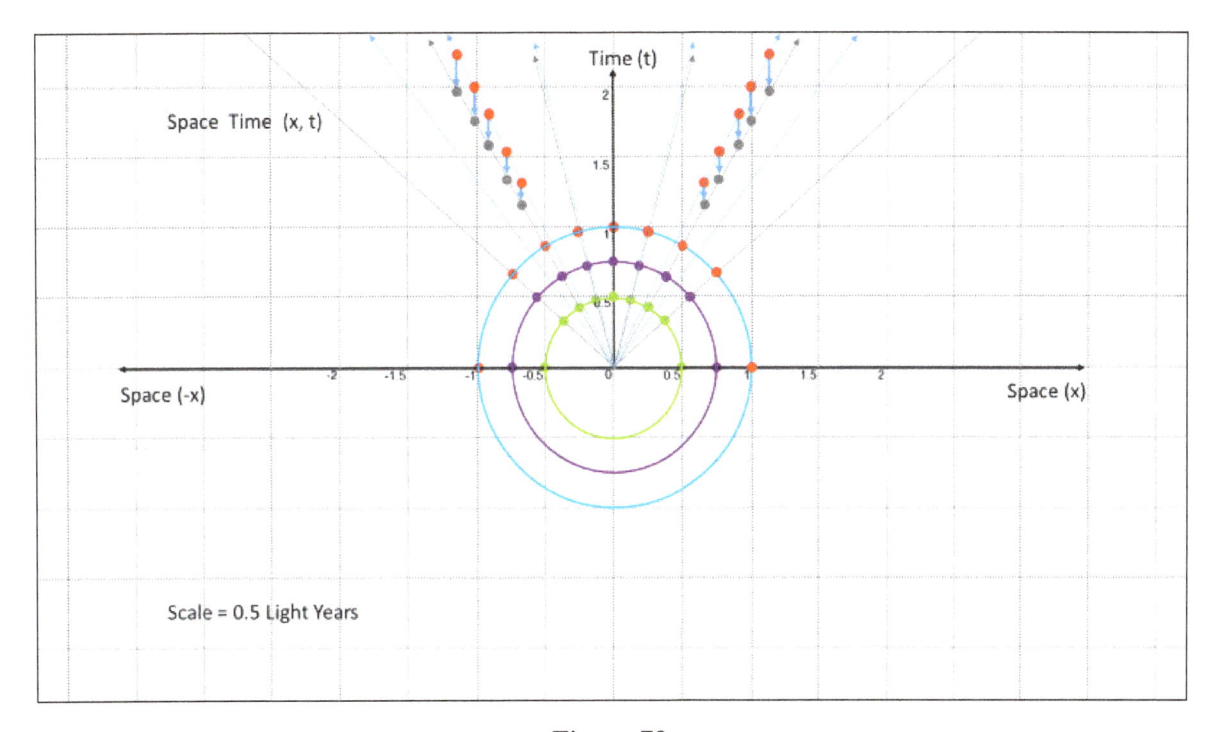

Figure 79.

Photons from more than a light-year away

What about light that is emitted from more than 1 light-year away? Do the rules about when we would see the photon still apply when the photon is emitted, at say, 1.5 (red dashed) and 2 (red) light-years away? We can see that they do. Our green stationary spaceship needs to wait 1.5 years to see the photon with the dashed red flight path and 2 years to see the photon with the red flight path. Our moving spaceships will encounter the photons along their time-dropped flight paths as before.

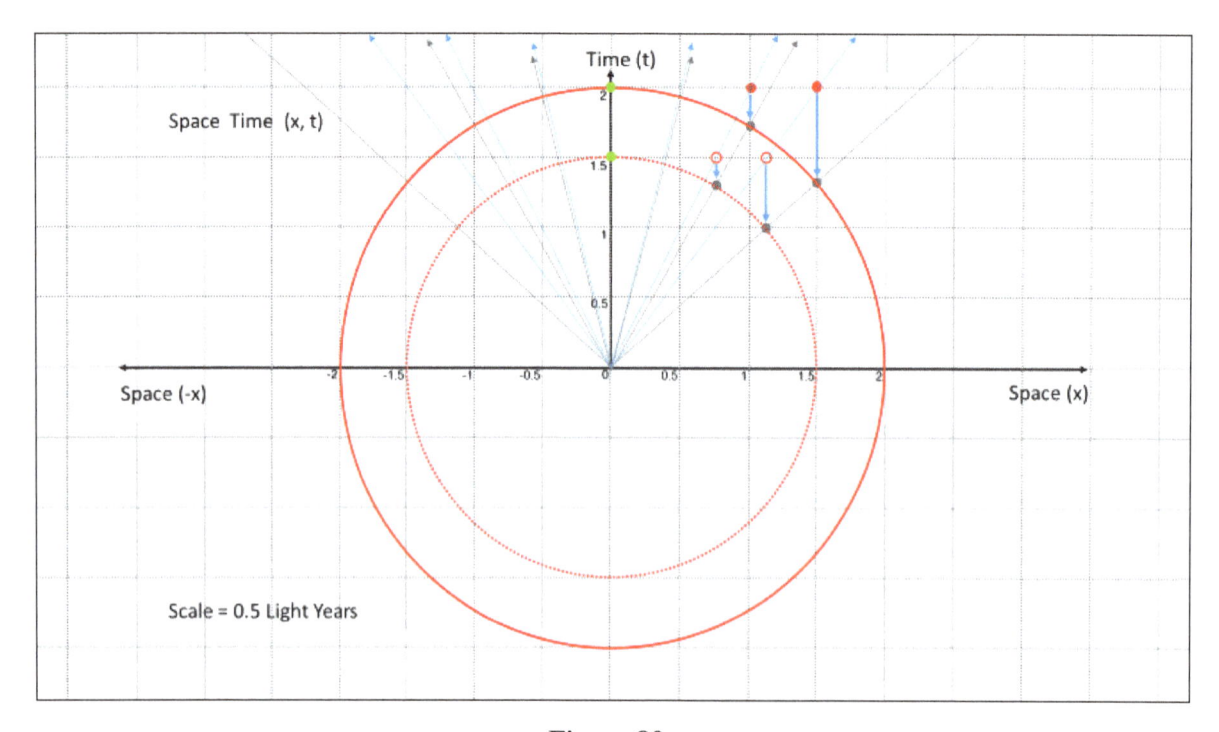

Figure 80.

What is fascinating here is that whatever the distance from a spaceship to the light source in light-years when the light is emitted, is exactly how many *years* will pass before the photon is encountered, regardless of the speed and direction of the spaceship. Not surprised? It's what the standard model tells us would happen. But here in the TRADD model, we can finally see how this seemingly counter intuitive result is achieved as a consequence of the path a photon takes through spacetime. Significantly, our new model tells us about the path through spacetime a photon *must* take in order for us to get the measurements we do. And as we can see – it's not a straight line!

Photon: It's coming – It's here – It's been and gone

OK. But what about a light source that is 1 light-year away and the photon has travelled for less than 1 year? We need to wait the full year before we can see it but how would it look in the graph after having travelled only half a year? And what about light that has been travelling for more than 1 year that we may have encountered after it had travelled a year. How would it look to us in the graph after a year if we didn't stop it when we could see it and let it travel on?

The graph below shows a light source 1 light-year away. It shows a photon that has been travelling for ½ a year (red), 1 year (blue), 1 ½ years (brown) and 2 years (green). The black arrows show how far up the time (t) axis the photon has gone to prove it has been travelling that long since leaving its starting point at t=0.

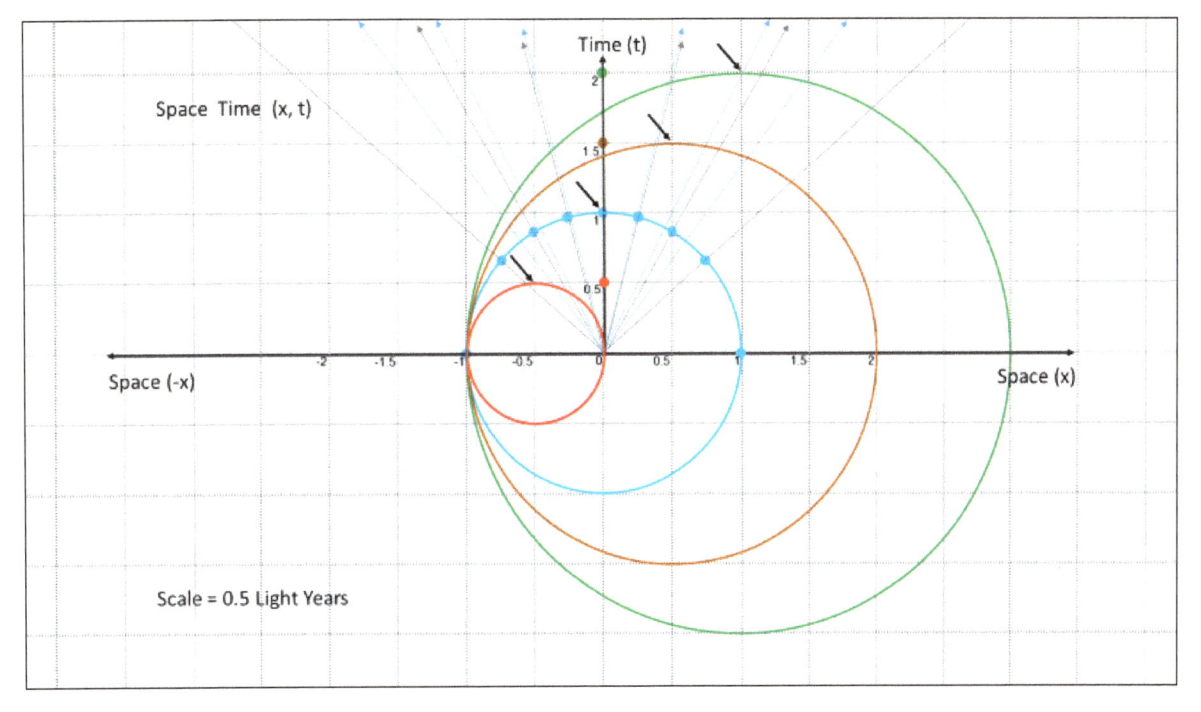

Figure 81.

Now, we need to look at our stationary spaceship at x=0. Let's get inside the spaceship. At t=0.5 we do not see the red photon. It has not arrived yet. We see the blue photon after waiting for a year, at t=1. After waiting for 1 ½ years, we do not see the brown photon. That's because it is really the same photon as the red and blue photon that has already been seen at t=1, when it was blue. So, when we are at t=1.5 the brown photon is behind us – in time. If we call the photon arriving at x=0 an event, then we can say at t=1.5 that the event has already happened. Indeed, it happened at t=1. Similarly, when we wait until t=2 for the green photon, it will not arrive. Again, because the event has already happened. It happened at t=1.

The graph shows the same photon having travelled for different lengths of time. It demonstrates that our stationary spaceship has only one opportunity to see it. That's when the photon and the spaceship are at the same place, at the same time. When they are at the same spacetime. Our spaceship cannot travel as fast as the photon. It cannot travel faster than the photon. And so, it will only ever have one opportunity to see the photon. It can never encounter the same photon more than once.

Photons that keep travelling

But that still leaves us with an unanswered question. Will our moving spaceships see the photon that was 1 light-year away from x=0, after the photon has travelled for more than a year? Remember that our spaceship flew through t=0 at x=0 so that at t=0 the photon was exactly 1 light-year away. The brown circle shows the flight path of the photon after it has travelled for 1 ½ years.

The grey marker represents the time-dropped location of our spaceship after travelling for 1 ½ years at ½ light-speed. It is at x=¾ along the x-axis.

To see the photon that has travelled for 1 ½ years it would be looking for a photon that is travelling at a speed of 1½ light-speed. This is because the spaceship is travelling at ½ light-speed and expects the approach speed of the photon to be light-speed, which is the speed of the photon minus the speed of the spaceship flying away from it – as before.

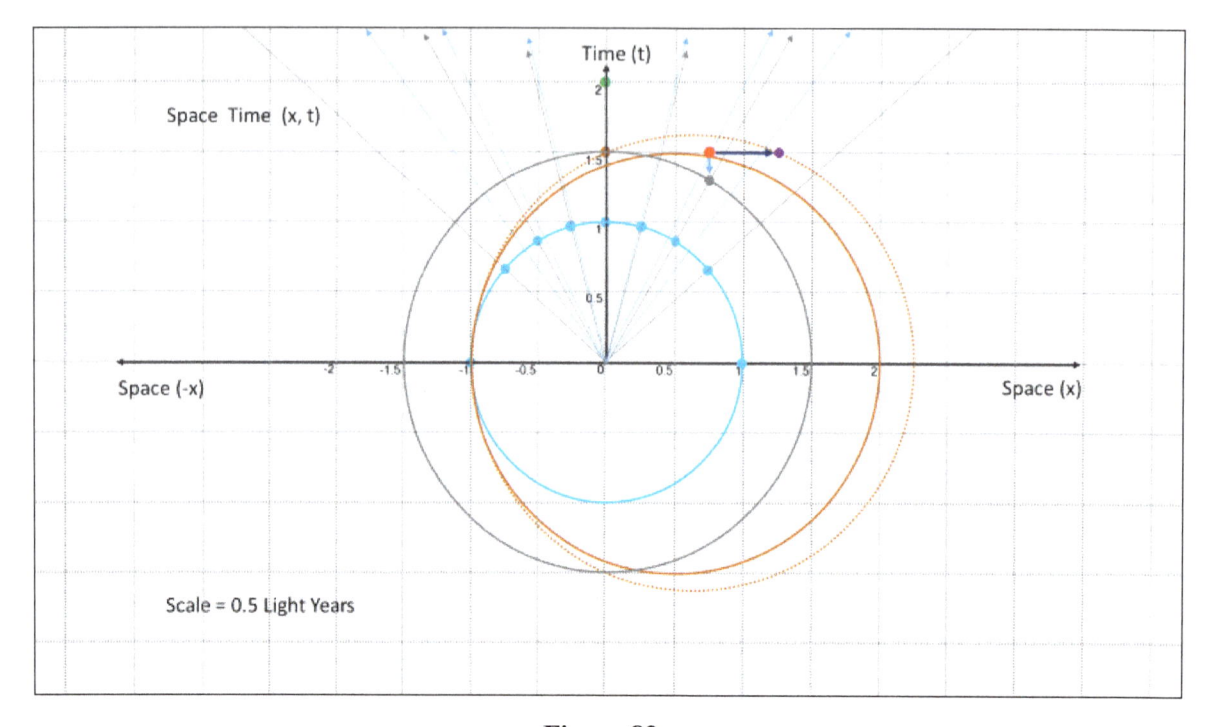

Figure 82.

For a photon to have a speed of 1 ½ light-speed and been travelling from x=-1 and t=0 for 1 ½ years we can calculate where it would be in x. At 1 ½ light-speed travelling for 1 ½ years would achieve a travelled distance of 2 ¼ light-years. From x=-1 that would get it to x=1 ¼. And we know that t=1 ½. So, we can place a purple marker and point to it with a black arrow to show how much further along x the photon has travelled from our spaceship at x=¾

What we find is that our spaceship could see the photon when it flew through the blue circle. That's when the photon had travelled for a year. But after that, the photon travelled on past in the x direction. So that if we could measure where it would be after 1 ½ years travel, we would find it to the right of us in x. In front of us. It would be at x=1¼ when we are at x=¾, half a light-year away.

Without using more circles and just to make sure we were not cheating, let's calculate where a photon would be, after traveling for 1 year so that it *can* be seen by our ship travelling away from it at half light-speed. The spaceship travelling at ½ light-speed would again be looking for a photon travelling at 1.5 light-speed

For a photon to have a speed of 1.5 light-speed and been travelling from x=-1 and t=0 for 1 year, we can calculate where it would be in x. At 1.5 light-speed travelling for 1 year would achieve a travelled distance of 1.5 light-years. From x=-1 that would get it to x=0.5 at t=1. So, we can place a purple marker at that point and point to it with a *red* arrow to make it easier to identify. We also see that it is at the same value of x as our time-dropped blue marker. And if we time-dropped the purple marker it would land on the position of the blue marker – exactly.

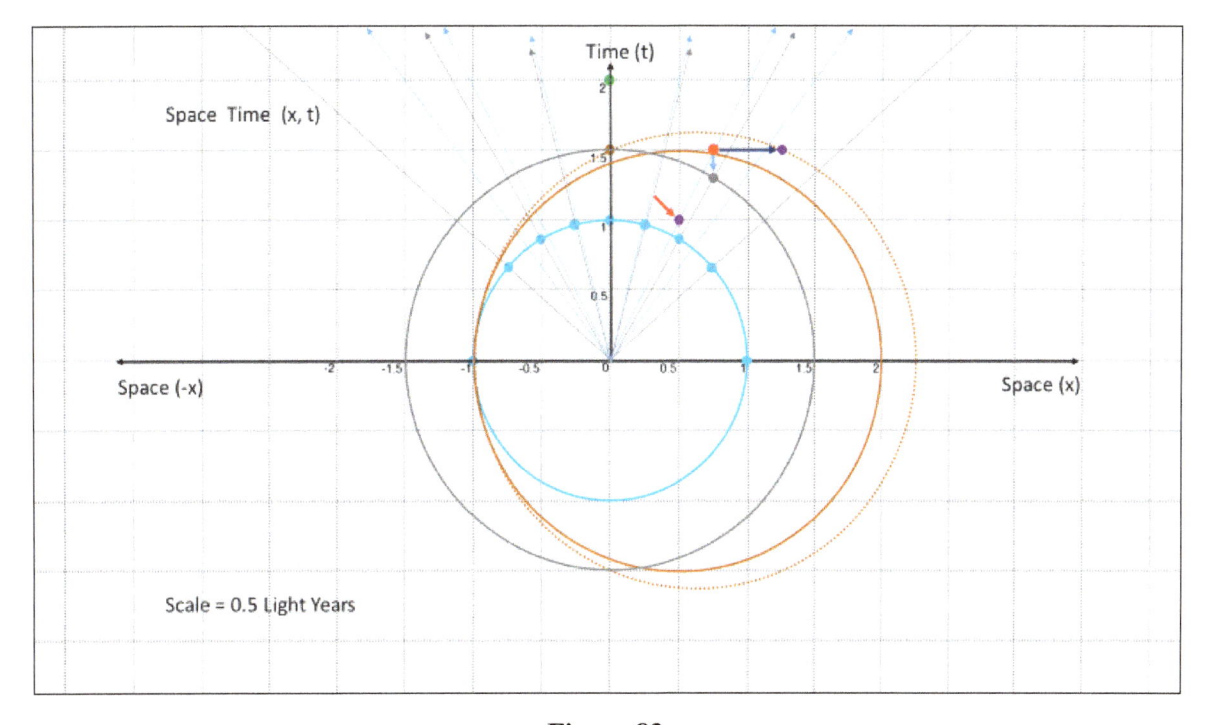

Figure 83.

143

Total Relativity and Dimensional Dynamics

Now, let's do one more, but this time for our spaceship travelling at ¾ light-speed. It will be looking for a photon travelling at 1¾ light-speed that has been travelling for 1½ years. We can do the calculation for x as before. But let's do it right here and break the calculation into parts so you can follow it. Travelling at 1.75 light-speed for 1 year will move the photon 1.75 light-years. Travelling at 1.75 light-speed for half a year would move the photon half of 1.75 light-years which comes to 0.875 light-years. If we add the two values together, we get $1.75 + 0.875 = 2.625$ or 2 and 5/8th light-years. So, from x=-1 that would get the photon to x=1.625 and t=1.5. So, we can again place a purple maker at x=1.625 and t=1.5 and a slightly shorter black arrow to show how much further in the x direction that photon would be from our spaceship travelling at ¾ light-speed.

What is interesting here, and expected, is that the photon sought by our spaceship travelling at ¾ light-speed after 1½ years is closer to it than the photon sought by our spaceship travelling at ½ light-speed. That's simply because it takes longer for the photon to open a distance between it and a faster spaceship than between it and a slower one.

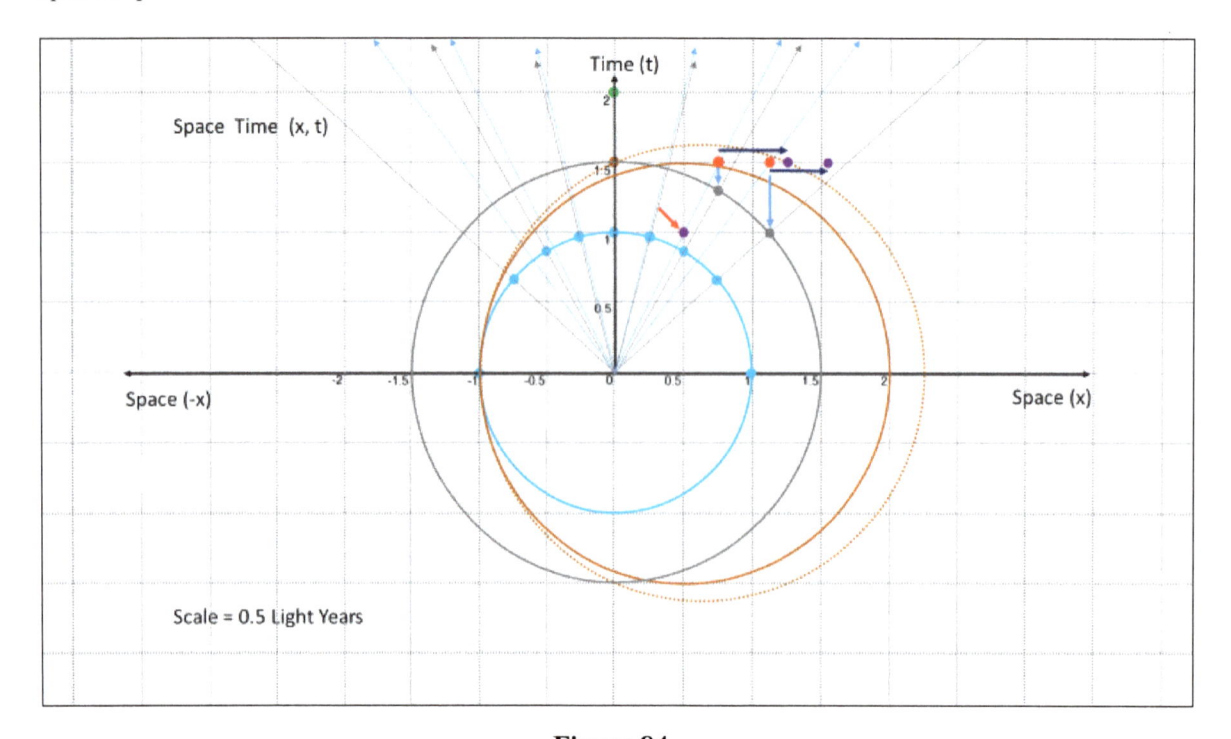

Figure 84.

What is light?

More importantly, we can now extrapolate these results to a spaceship travelling at any speed. So, we know that even a spaceship travelling at, say, 0.1 light-speed (10%) will find the photon a further distance along x than the spaceship is when it makes the calculation. The calculation is 1.1 x 1.5 = 1.65 and so the photon sought would be at x=0.65 and t=1.5. The graph uses a red arrow for our spaceship travelling at 10% light-speed.

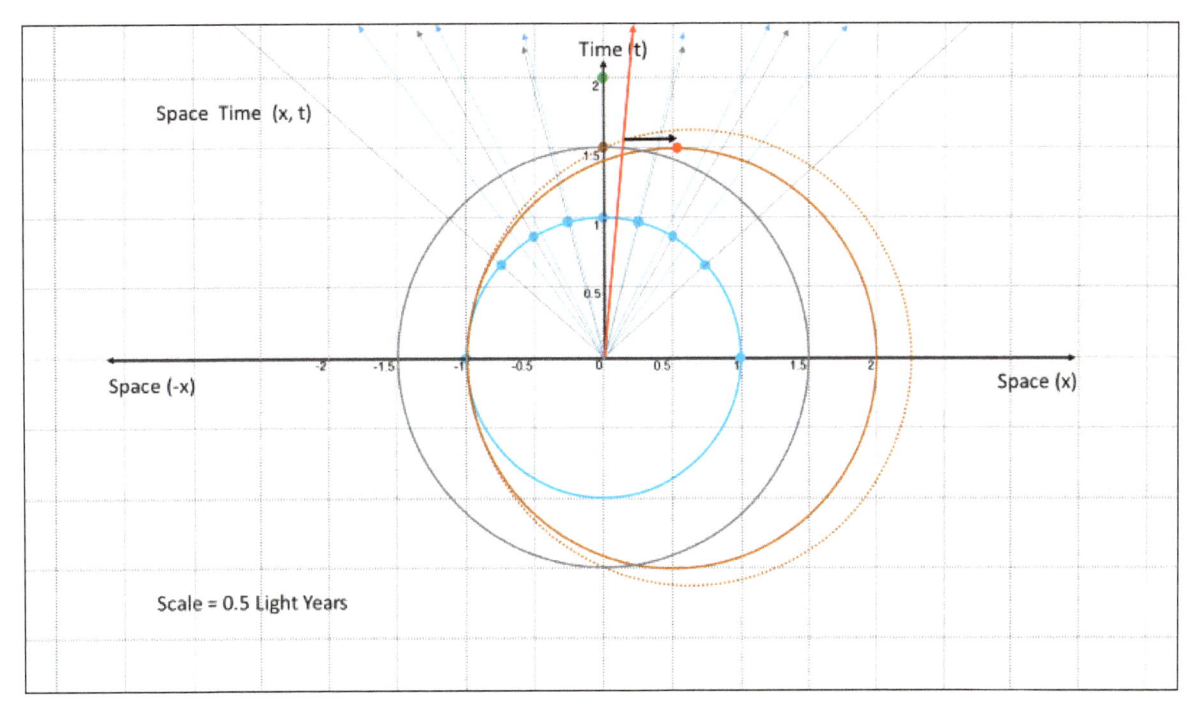

Figure 85.

For a spaceship travelling at 1% of light-speed the calculation is 1.01 x 1.5 = 1.515 and so the photon sought would be at x=0.515 and t=1.5. The graph uses a red arrow for our spaceship travelling at 1% light-speed.

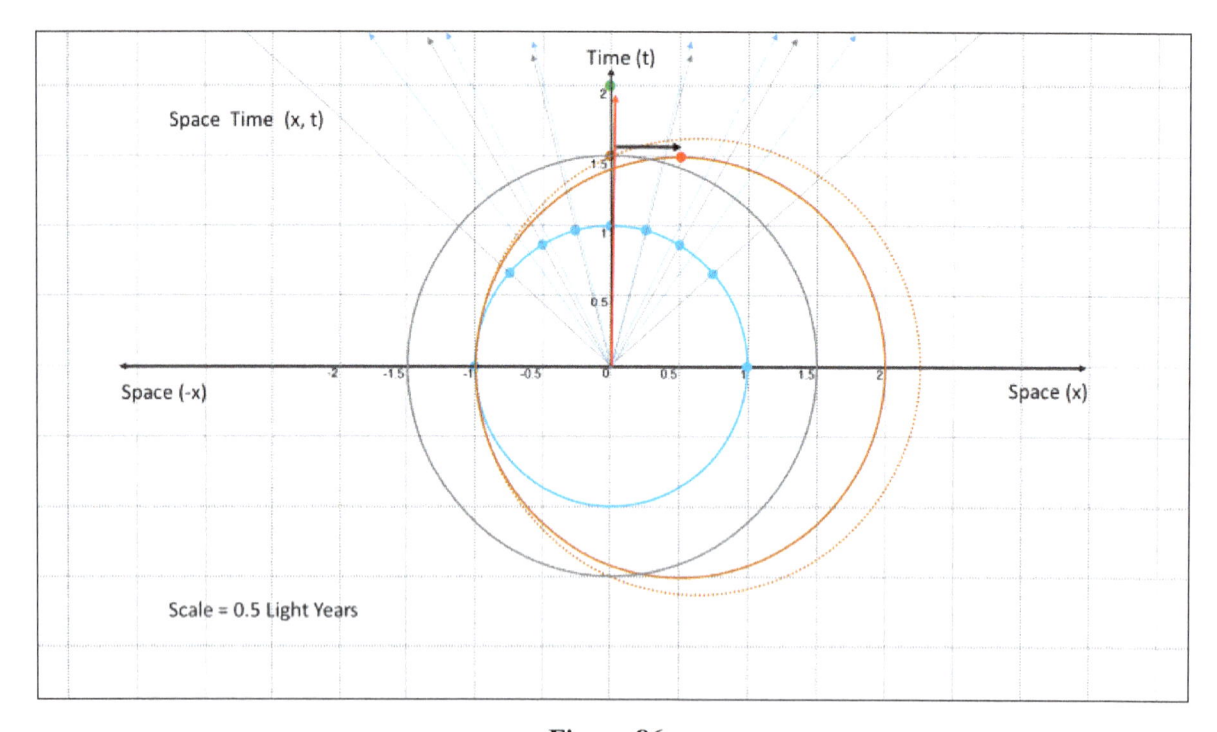

Figure 86.

If you look carefully, you will see that the black arrow from our ship travelling at 1% light-speed is longer than the one from our ship travelling at 10% light-speed.

What is light?

Faster through space than time

We can see that after travelling ½ a year the red photon has travelled 1 light-year when we measure the distance between x=-1 and x= 0. After travelling 1 year, the blue photon has travelled 2 light-years. After travelling 1 ½ years, the brown photon has travelled 3 light-years. And finally, after traveling for 2 years, the green photon has travelled 4 light-years when we measure the distance between x=-1 and x=3. The distances we are measuring here, represent the diameters of the circles – twice the radius.

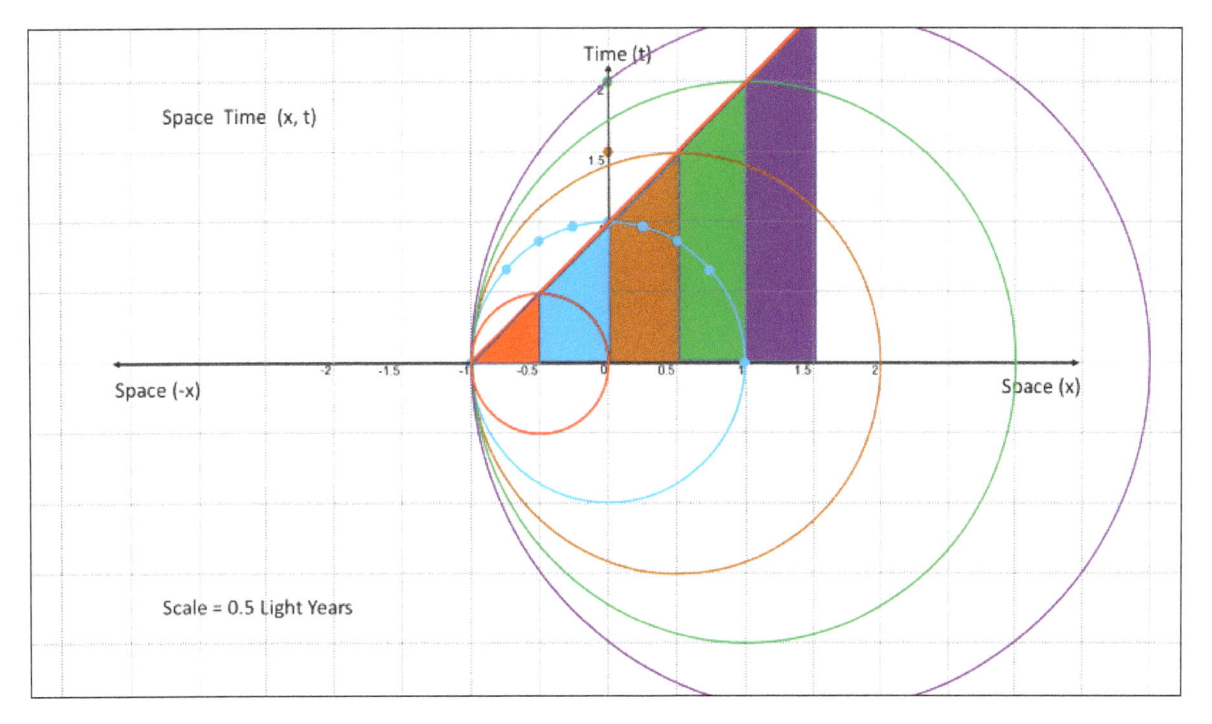

Figure 87.

We see the flight path of the photon changes as it moves and according to how long it has been travelling. We can see that it progresses faster in the x direction than in the time direction. For every light-year travelled in the x direction it travels only half a year in the t direction. When it is at t=2 (the green circle) it is also at x=3 which from its starting point at x=-1 is 4 light-years. So, in 2 years it has travelled 4 light-years. It is important to recognise that this is also a consequence of the scale we have used.

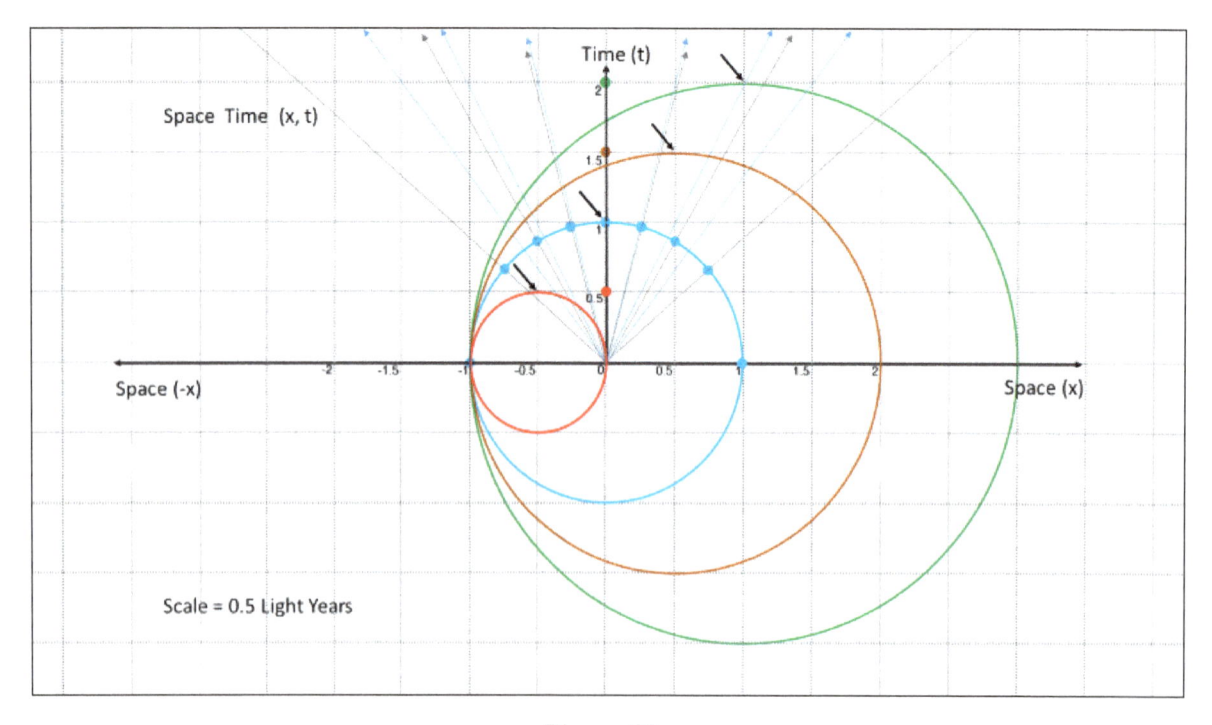

Figure 88.

Light is energy that travels as a wave

We started with the idea that the approach speed of light, when measured, is always the same – it's a 'constant' value. And we set out to understand how that could be.

We have shown that a photon from a source 1 light-year away, will be encountered after exactly 1 year by our stationary spaceship. And from a light source 2 light-years away, will be encountered after exactly two years by our stationary spaceship. However many light-years away from the light source, that's how many years our stationary spaceship must wait to encounter it.

We have shown that the approach speed of the photon is always light-speed, regardless of the direction or speed of our spaceship. And we have shown that the approach speed is light-speed regardless of how far away the source of the photon is.

But we *knew* the approach speed was constant. What we have shown is *how* an approach speed could be measured as *constant* and then draw some conclusions about what that must mean. The result seems to indicate that light travels as a wave that propagates by extending its radius at a constant speed, that is, light-speed. That means the radius of the wave increases by 1 light-year every year. So we can say that the speed of light is *constant* if we say that light travels as a *wave*.

If you take the centre of the circle as starting out at x=-1 and t=0, the light source, we see the centre of the circle travel in the x-direction at a speed of 1 light-year per year. But as the centre of the circle moves off in the x-direction, the wave path of the photon spreads out. And it spreads out by extending its radius at a rate of 1 light-year per year. So two motions occur simultaneously. The *point* in spacetime where light is emitted travels through space in the x-direction at light-speed. Significantly, it does not travel through time. But light itself, radiates outward, as a wave, from that (travelling) point, also at light-speed.

A consequence of this dynamic is that whenever the approach speed of light is measured, the value is always the same – light-speed. If light were a particle (not travelling as wave), this result would elude us; its approach speed could never be constant, but would vary. Notice that we are not saying that light is a wave. We are saying that it travels as a wave.

So how might we measure it as a particle? Ever been swimming in the sea when there are waves rolling into shore. You go up and down with the wave as you're swimming. You are not 'hit' by the wave. You don't feel its energy. But now stand in the shallows and plant your feet in the sand of the seabed. When the wave arrives – you can feel the force of it. Can light somehow propagate as a wave which, just as a water wave, imparts force when it collides with something? And when it imparts force, do we then measure it as a particle?

The other fascinating observation from the model is that if we were to animate the movement of the emission point and of the radiation of the wave, we will see the photon always approaches the spaceships from below. In other words, light travels through spacetime as a wave, but *approaches* us through time, not through space. And more than that, it approaches us from the past, not from the future – and that's appropriate. We might have notions about light travelling to us through space, but it doesn't. Not really. Whilst it travels through spacetime as a wave, we only see it when we see it. We never ever see it *coming* toward us. We can't because it's always remote to us in time. And because we see a photon when it's absorbed. We think we can see it coming when we see a beam of light, like when we shine a torch at night. But that's an illusion. What we see are photons *from* the beam, popping out of the past, into the present, and presenting to our eyes or our instruments.

The whole picture

You know I've left out a large chunk of information and some of you may have already figured out what that is. But I had to leave it out until now, otherwise charts become so busy that we want to turn our eyes away from them and turn our minds off. You see, the light source we used was directing photons in the positive direction of the x-axis. The light was shining from left to right. But when light is emitted, at least in this TRADD model, it spreads out in all directions. It must. That's because the model finds that light does not have its own velocity. It moves by being carried by the expansion of space. And in the model, space expands in all directions. Indeed spacetime appears to expand in all directions, because the expanding xyz-volume is moving through time as it expands. And although we used a slice of spacetime in the xw-plane, which we later renamed the xt-plane, which we later renamed spacetime – a slice of spacetime – we did that to make it easier to see what happens when we move a volume in x, y, z, in the w-direction (in the time-direction).

What is time?

Time is how we perceive travelling through a fourth spatial dimension.

If you've jumped straight to this section without first reading the TRADD perspective on space, gravity, and light, then you'll likely not have a clue as to what I'm talking about.

In the section on space, we gave space *attributes* to render it *something* rather than *nothing*. We then cast our three-dimensional universe as the boundary of a hyperspace (a four-dimensional space). We did away with the spacetime continuum by doing away with time. That left us with space – just space – three-dimensional space.

The section on gravity introduced time as *what we perceive* when our universe, all three spatial dimensions of it, travel in the w-direction. We therefore reinstated time as the units along the w-axis and renamed it the time-axis. Yea, we did that. We got rid of time. Then brought it back! We showed that if our, now three-dimensional universe, travelled in the w-direction (renamed the time-direction), we could monitor that progress in any of the x, y, or z axes. We showed why progress was slower in the vicinity of mass than remote to it, and how mass contracts space, which we measure as gravity. Later, we'll show why progress is slower in the w-direction when we move through space at speed. Any speed. We identified the progression of the x, y, and z axes, in the w-direction, as the mechanism responsible for the expansion of space and for the phenomenon we experience and measure as time. We showed how due to its movement *through* time, space consequently expands, at n-metres per year, and that 'n' is constant. The expansion of space underwrites change, rendering a dynamic universe. That this mechanism operates everywhere at a constant rate, relating distance and time through the constant 'n', offers an enticing explanation for the observed consistency in the mechanics of space and time – why the physical laws of the universe are everywhere the same.

The section on light models how light might travel so that when its approach speed is measured, we always get the same value. It spreads out as a wave through space and time, extending its radius by 1 light-year every year whilst the origin of the radius travels in the direction of motion at a speed of 1 light-year per year. In other words, its radius of propagation is a function of time. By extension, it is also a function of expanding space. Significantly, it spreads out in all four spatial directions: x, y, z, and w (time). We show that time varies from place to place, due to the presence of mass. Later, we show that time varies from place to place due to changing proximities as we travel – as we move through space. So, too, must the expansion of space vary from place to place when measured as n-metres per year, because *per year* means an elapse time of 1 year, and that elapse time varies (from place to place). Hence, the expansion rate of space with respect to time is the same everywhere. Time and space are thus inextricably linked. For a light wave to spread out by extending its radius by 1 light-year every year, it too must experience time varying from place to place, and distances varying from place to place due to spatial expansion and contraction. To complete any journey spreading out at a rate of 1 light-year per year, a light wave would need to move in concert with spatial expansion – with the expansion of space. To do that, it would need to move *with* space rather than *through* space. As space expands, light moves. We conclude that light has no speed. No *inherent* speed. Like driftwood in a current, or dust in the wind, light just goes along for the ride. It spreads out as space spreads out – as space expands, at n-metres per year. We have a measurement for that. We call it light-speed or C. But under the TRADD model, we find C is the rate at which space expands. Hence, *n-metres per year* becomes *C-metres per year*. A constant speed that makes sense. The speed of space.

TRADD suggests that time is nothing more than how we perceive ourselves moving in the w-direction. The mechanics of that movement, however, expands space. The expansion of space manifests change; the transition from one state to another. This leads us to explore time in more detail. What are the characteristics of time? And how does our new model accommodate these characteristics?

The characteristics of time

In the section on gravity, we did away with the spacetime continuum. We did this to remove 'Time', leaving three spatial dimensions. We then introduced a fourth spatial dimension and named it the w-direction. Each of our four spatial dimensions could be scaled as distances. We then proposed that the three spatial directions, x, y, and z, were all at once moving in the w-direction. We suggested that we sense that movement as the passing of time. We next suggested that such movement caused spatial expansion in the x, y, and z directions, at a rate of n-metres per year. We then suggested that the expansion manifests state changes. We then came full circle to suggest that for these reasons, we associate *change* with the perceived passing of time.

Time is a direction

In the TRADD model, time is fundamentally a direction. TRADD initially models our universe as the three-dimensional boundary of a hyperspace – a four-dimensional space. After assigning attributes to space in general, it demonstrates that three-dimensional space can be the boundary of a hyperspace in the same way that a two-dimensional space, a surface area, can be the boundary of a three-dimensional space, a volume. It demonstrates how the deformation of hyperspace causes its boundary to expand. That happens when the boundary is pushed or pulled in a direction perpendicular to it. When modelling hyperspace, we called that perpendicular direction the w-direction. The axis of the w-direction had units of length. It was only after suggesting we might perceive movement in the w-direction as time, that we rescaled the w-axis. We changed its units from length to time.

We perceive time because we move through it

Having rescaled the w-axis to units of time, it made sense to refer to it as the time-axis, pointing in the time-*direction*. So, whereas we initially spoke about moving in the w-direction, we subsequently spoke about moving in the time-direction. We explained that it's the same thing. It doesn't matter what units you use for the axis.

But one differentiator was the operation itself, of moving *through* time as opposed to the notion of time passing. While time is how we perceive movement in the time-direction, it is that *perception* that leads us to talk in terms of time passing, the flow of time, the passage of time and so on. Our perception of time extends to notions of past, present, and future. And to concepts like before, after, and duration. It's as though time has been adopted as an entity having attributes that characterise it. It flows, but only in a forward direction. It passes, but only in the present, and only from the future to the present to the past – in a backward direction. It forms part of the spacetime continuum that deforms in the presence of mass. It progresses more slowly in a gravitational field. And it passes progressively more slowly the faster we travel. These last two attributes we call time dilation – the slowing down of time.

By having the perception of time and by seeing events unfold, we have made the connection between the perceived passage of time and the changes we see around us. Where TRADD says that as we move in the time direction, space expands, manifesting change – TRADD asks us to interpret that to mean that because we move *through* time, things change. Instead, we do it the other way around. We pay attention to change whilst noting the passing of time and employ the perception of time to measure the duration between events manifesting the change. In other words, we use change to calibrate the passing of time, whereas TRADD ask us to use time to calibrate change.

We even found ways to measure the passing of time using all manner of devices that we collectively call clocks. These devices all operate by measuring change – the very change that the progression *through* time manifests indirectly via the spatial expansion that progression initiates.

I don't mean to be critical here. But it's important to contrast the new TRADD model with the standard model. Science has done a superb job with time. We can measure it with astonishing precision. But where doing that deals with measuring a perception of what we call time, TRADD would say we are measuring change brought about by our progression *through* time. The difference is subtle enough and some might say trivial. But as we shall see, treating that difference as significant, helps us understand some of the stranger characteristics of time. It brings an understanding that has hitherto eluded us.

We travel through time at light-speed

In so much as light-speed can be defined as 1 light-year of distance travelled in 1 year, TRADD identifies the speed with which we move through time to be just that – light-speed. It's an irony to talk about travelling through time at some speed – any speed – since speed is notionally calculated as the distance travelled per the time taken to travel it. But here we are – moving through time – at speed!

TRADD comes to this speed – light-speed – from the perspective of the units used to scale the axes that define each of the four spatial dimensions of the hyperspace it claims our universe is bounding. When we labelled the directions we named their axes as x, y, z, and w. And because these are directions, we scale them in length units – in metres. When we renamed the w-axis to the time-axis, we scaled it in time units – in years.

Doing that did not automatically equate 1 light-year in a directional axis to 1 year in the time-axis. Although for aesthetics, that was implied by the scale used in our charts. It was the emergence of the circular wave front representing the motion of light through spacetime that tipped us off. When the wave front had travelled 1 year, its wave had extended its radius by 1 light-year. Sure, we used the fact that the approach speed of light is measured to be 1 light-year per year. But that wasn't enough. We also had to use the fact that the faster one travels through space, the slower time flows – or in the TRADD context, the slower we move *through* time. Taken together, these measurements, forced you might say, for the time scale of 1 year to equate to the distance scale of 1 light-year. And as we will see later – it can't be otherwise.

Hence, travelling a measure of 1 year up the time-axis equates to travelling 1 light-year along a length axis. And so our speed when travelling *through* time is always light-speed. It is always 1 light-year per year. Well, almost always – but we'll get to that later.

When we travel through time, space expands

As our three-dimensional universe moves through time, it expands – it stretches. There is a significance about this dynamic that relates to the so called 'speed of light'. For light to move through space (and time) as we have modelled it, and for our measurement of its approach speed to always be the same, there is one other factor that emerges as consequential. The light wave must move in concert with the expanding space.

Locally, each year, the surrounding space *stretches* at a rate of n-metres per year. That means that as time progresses, light from a distant source would need to travel further to reach our locality. On arrival we measure its approach speed as 1 light-year per year. And there's the hint.

Locally, 1 year is just that – 1 year. But between the light source and our locality, there may be numerous other localities in space where due to the presence of mass, say, the progression of a year is faster or slower from one locality to the next. But we measure the speed of light according to a year at our locality. We measure it using local time. And we know that at whatever other locality you might be between our locality and the source, you would there, also, measure light to have the same speed, even as the elapse time of a year in each locality may be different – in relative terms. So in each locality, the speed of light is measured according to the progression of 1 year *at that locality*.

So we know that time progresses at varying rates from place to place, and yet, the speed of light appears to be tied everywhere to whatever the rate happens to be at that location. And that its *speed* remains 1 light-year per year according to the elapse time of a year at whatever location of space it happens to be passing through, one might argue that somehow light knows how to adjust its distance travelled, so as not to travel more or less than 1 light-year during the local elapse time of 1 year. In other words, we would need to accept that somehow light knows when to speed up or slow down to local time, and then exactly traverse 1 light-year during the elapse of the destination's local time for a year. But that's still not going to be enough for it to be measured everywhere at a constant speed. It won't work.

As time elapses, space expands. So what does that mean for light travelling through different localities having different elapse times for a year. Let's look at light travelling from a source to our locality but traversing a locality along the way where elapse time is much slower, due to lots of mass, say, and hence a strong gravitational field. If we say the distance between the source and our location is exactly 1 light-year, then we would expect light to arrive at our location exactly 1 year after transmission, and according to our elapse time of that year locally. The problem is that whilst the light is travelling, space continues to expand. It doesn't expand uniformly (in our TRADD model), but variously and according to the elapse of time at different locations along its journey.

So, for light from 1 light-year away to arrive at our location after exactly 1 year, it needs to do several things. As it travels it needs to continually honour local elapse time by not exceeding a speed of 1 light-year per local elapse time year. And it can't go slower than that speed, either. Then also, it needs to ensure that it arrives at our location exactly after 1 year has elapsed locally at our location but take a little longer than 1 year to account for the increased travel distance due to space having expanded along the route. Now that's a lot to ask of a wave travelling through space. We'd be asking that the light wave exercise a level of intelligence or at least that an intelligent mechanism is at play. At all times. It just can't happen that way!

But if light hitched a ride with the expanding space, that space would extend from the light source to us at a rate of exactly 1 light-year per year. The speed of space! That's per year based on our local elapsed time. And that's indeed what we measure the approach speed of light to be. And at any location in space that's how the approach speed of light will be measured – but only if it hitched a ride on the mechanism that is constant with respect to time – the expansion rate of space.

Expanding space manifests change

If we recall why in the TRADD model space expands, we said it was because our three-dimensional universe was moving through time. And as the bounding volume of a hyperspace, as it moves through time (in the w-direction), it expands. But we also introduced forces to show that the expansion of space occurs at a rate, governed by competing forces. Force implies energy and changes in forces implies the transformation of energy. So, as space expands, changes occur in the way these forces are balanced – or rather, the imbalance in these competing forces generate change. Hence three changes occur simultaneously. The progression of space, our universe, through time, in the w-direction. An increase in the volume of space, our universe, due to space expanding. Energy transformations, in the energies bound up in the forces that push space to expand and resist that expansion.

All *that* in a universe before we introduce mass. But the equivalence of mass and energy is noted in the standard model. Concentrate sufficient energy into a volume and perhaps ... just maybe ... you end up with mass. Conversely, break mass apart and the energy bound up in it is dispersed – or as we like to say, released. The pushing and pulling forces that hold our space together represent a tremendous amount of energy. And as space is stretched, that energy is transformed.

As we move through time, the status of competing forces and their energies change along with the change in the amount of space there is. These changes transform our universe from one state to another. These fundamental *state changes*, occurring continually as we move through time, manifest to us as … change happening as time passes.

Change is consistent with the 'flow' of time

We have shown that changes that occur in our universe as it moves through time, happen continually, due to the expansion of space that moving through time causes. But we have also explained that the pace with which we move through time varies from one location to the next, according to the energy density per unit volume. Of course, we used a simplified example to demonstrate this in the model. We increased the energy density in space along the x-axis by introducing mass. Mass represents a concentration of energy. We showed *how* where energy (mass) is concentrated, the progression of space through time happens at a slower rate than remote to that energy concentration.

And because change in the universe – the state changes we talked about – happen due to the movement *through* time and the associated expansion of space, change remains consistent with what we perceive as the *flow* of time. Hence, wherever you might be in the universe, the rate of change, the speed with which things happen, the elapsed time between one event and another – remains the same. We experience the universe the same way where time is flowing faster as where time is flowing slower. It's constant. So a clock based on the counting of events made in one part of the universe where the energy density is low, in a weak gravitational field say, would work the same as a clock made in a different part of the universe where the energy density is higher. In a stronger gravitational field.

It is this consistency, innate as the rate of change, that is experienced when we move one of two clocks from a strong gravitational field to a weaker gravitational field. The clock in the strong gravitational field will travel through time at a slower pace than the clock moved to the weaker gravitational field – it will travel in the time direction at a slower pace. Each clock ticks off seconds consistent with the pace with which it moves upward along the time-axis. And the expansion rate of space – the pace of change – keeps the *rate* of change (the elapsed time between state changes), consistent with the *flow* of time. So the clock travelling more slowly up the time-axis ticks at the *same* rate *relative to the space around it* as does the clock moving more quickly up the time-axis. The faster travelling clock – travelling through time – will tick faster relative to the slower travelling clock, but each clock will keep the correct time locally, consistent with its pace up the time-axis, because the rate of change, which the clock is counting, is tied to the expansion of space, which in turn is tied to progress along the time-axis.

When the clocks are, later, brought together again, to the same position along the time-axis in space, and hence to a place where the gravitational field is the same strength for both clocks, they will tick at the same rate again *relative to each other*. But the clock that remained in the stronger gravitational field for a while, will have lost time. It lost time because it ticked more slowly than the other clock for the *duration* that the clocks were separated.

Here, the term *duration* is also relative. How long were the clocks separated along the time-axis? If we say both clocks started out at the base of a mountain having a stronger gravitational field than at its peak, we would have moved one of the clocks to the peak, for a while – into the weaker gravitational field. *For a duration. For a while* represents the duration. If we used the clock left at the base of the mountain to time how long the other clock was at the peak, we might get a value of 1 year, say. But that's 1 year according to the clock that ticked more slowly at the base of the mountain. If instead we used the clock that we moved to the peak of the mountain to time how long it was up there, we would get a value of more than 1 year. That's because the clock at the peak ticked more quickly than the clock at the base, all while the clock at the base measured off 1 year.

How time dilation works within the TRADD model is considerably different from how it is presented as working in the standard model. That's because the standard model differs in its description of gravity and is almost void of a description for time. In the standard model, it is the curvature of space that lengthens geodesics because the curvature stretches spacetime, compelling objects to take a longer path through the space as compared to where space is less curved in a weaker gravitational field. Taking a longer path takes *more time*, it's slower. But that's not the whole story.

Take two clocks, travelling at the same speed over the same distance, but where one takes a longer path due to the curvature of spacetime – a path through a stronger gravitational field. This is analogous to them travelling from one side of a circle to the other – the same point to point distance, left to right. The fast route is along the diameter and the slow route is along the circumference. When the fast route clock is halfway through its journey, at the centre of the circle, the slow route clock is not yet one quarter way around the circumference (which would be its halfway point). The analogy would say the clock on the circumference is travelling more slowly from left to right. Both clocks are travelling at the same speed, but time along the slow route must be progressing more slowly.

Geometrically, because it is spacetime that is curved, not only is the geodesic path through space longer (stretched), but the path through time is longer too. So, we have time stretched out along the circumference of the circle, not only space, being the *length* needed to travel to the other side. So travelling through a region of spacetime where time is stretched out, causes time to pass more slowly as compared to where it is less stretched out. Spacetime is more curved the stronger the gravitational field. In curved spacetime, both space *and time* are stretched out into a longer pathed geodesic. So, it is not only the distance through space that is longer, but the distance through time is longer too – as compared to a region having a weaker gravitational field. Wherever time is stretched out in a region of space, clocks tick slower than in regions where it is less stretched out.

The mechanics of spacetime

The mechanics of spacetime is really the mechanics of space. In the section on space, we asked why if space was expanding, it didn't just expand in an instant. And *that* it didn't, implied a resistance to that happening. And resistance implies force. And force implies energy. So if space is expanding, then it does so at a rate.

In developing a model to show this, we noted the spatial directions comprising a four-dimensional construct having a three-dimensional boundary. With the directions x, y, and z, comprising the boundary, the w-direction perpendicular to these, pointed into the hyperspace. So as the boundary moves in the w-direction, it expands. It stretches out in the x, y, and z, directions. An expanding spatial boundary due to its movement. Significantly, the expansion is not only due to the movement, but proportional to it. It is the mechanism underlying spatial expansion. The mechanics of space.

If we say that this movement in the w-direction is how we experience time, then we can rename the w-axis the time-axis and rescale it from length units to time units – from light-years to years. What was a model having four spatial directions is now a model having three spatial directions and one time direction. So where we began looking at how space expands in relation to space, we ended up looking at how space expands in relation to time. So the mechanics of space *became* the mechanics of spacetime.

Accommodating time

So far, we have defined time. Fitted time to the TRADD model as a direction scaled in years. Demonstrated *how* we move *through* time at light-speed. Used time as a metric when exploring *how* light might travel. And when exploring gravity, we showed *how* time might slow down in the proximity of mass – time dilation in a gravitational field.

But we need to do more.

We need to accommodate time.

With our new definition for time and its characteristics, we need to explain *how* our measurements *of* time and our measurements *using* time produce the results they do – in a way that makes sense.

One of the most bizarre outcomes of the standard model is the so-called slowing down of time. Called time dilation, it is important to recognise that the standard model identifies two circumstances where this occurs. It occurs in a gravitational field. And it occurs when we speed up.

It confounds me that this does not get more attention. I don't find it especially strange that we simply accept the slowing down of time because that's what the standard model tells us and because we have validated it experimentally. But really? I mean, that time *slows down* – it's a big deal! Right?

And yet if you ask of the standard model – what is time? – well, there are opinions but very few answers. But … we have a hint. That time slows down under two 'seemingly' unrelated circumstances, should ring alarm bells for anyone asking the question. Could it be that time slowing down in the proximity of mass, and time slowing down the faster we travel, has something to do with the nature of time itself?

If the TRADD definition of time is a good one, that it is a direction, then surely the TRADD model ought to be able to offer up a sensible explanation for *how* and *why* time slows down – in *both* circumstances.

We have already dealt with time slowing down in the proximity of mass. To deal with the slowing down of time as we go faster, we first need to look at what that means. What do we mean when we say we go faster?

Velocity

We all know what 'speed' is – and some of us have paid speeding fines to prove it (or to prove that 'no, I didn't know what speed was ... but I do now'). Another term, closely related to speed is *velocity*. The difference between speed and velocity is taught in schools and most of us may remember that. It's important here, so for those who don't remember I'll do my best to explain it.

Speed does not consider the direction of travel whereas velocity does. Yes, it's that simple. So we can talk about speed in the first instance.

We say that if you travel seven kilometres during your morning run and it took an hour to do that, then your *average speed* is seven kilometres per hour. Of course, if you were able to maintain a consistent running pace, then at any point along the way, your *speed* would be seven kilometres per hour. But maintaining a constant speed when running is difficult, particularly if along that seven-kilometre stretch there are hills to climb, traffic lights to obey, or roads to cross – carefully. Any number of circumstances might interrupt your steady pace. That's why after we have travelled the distance and look at our watch, we can say that our average speed was 7 km/hr. Speed, then, is the distance travelled divided by the time it took to complete the distance.

Let's now look at *velocity*. If we add 'direction' and say that you travelled only in one direction and that direction was north, then we can say that your velocity is 7 km per hour, *north*. But what if we travelled to a different destination along a different route? If we travelled 3 km in a northerly direction and then 4 km in an easterly direction, we would arrive at a destination that is 3 km north of our starting point and then 4 km east from there. We would still have travelled 7 km and it would still have taken an hour. So our *speed* would still be 7 km per hour. But not our velocity.

If there was a road that could take us in an approximate northeast direction, we could have taken that road and arrived at our destination sooner. That road would be the long edge of a triangle. Going north for 3 km makes the first edge of the triangle – we can call that length 'a'. Then going east for 4 km makes the second edge of the triangle and gets us to our destination – we can call that length 'b'. If we join our starting point to our destination, that will complete the triangle – and we can call that length 'c'.

We can calculate the length of 'c'. It's 5 km.

We therefore only need to run at 5 km/hr to reach our destination in an hour if we travelled that road along the diagonal 'c'. If we wanted to run at an average speed of 7 km/hr, we would only have taken five sevenths of an hour or about 0.71 hours to reach our destination (around 42 minutes and 51 seconds).

What we learn from this is that speed is a measure of how fast you are travelling, stride by stride in your run, which at the end of your run, allows you to calculate your *average* speed as simply the distance travelled divided by the time taken. But average *velocity* ignores your stride-by-stride speed *and* your distance travelled. Because velocity takes notice of direction, it will only consider the direction from your starting point to your destination, and regard your distance travelled as the straight line along *that* direction between those two points. It then uses *that* distance travelled divided by the time taken to give you the average speed travelled in *that* direction – and a speed travelled in a particular direction, we call *velocity*.

Our velocity through spacetime

I would like you to look around and find a stationary object; something you might pick up and take with you or pick up to use. It can be a cup of coffee – as an example. Now look at that object. You can see it is not moving. You might argue that it is moving because the earth is spinning on its axis and in addition to revolving around the Sun, our Sun is moving around the galaxy and the galaxy is moving through space. That's fine. If you say it is moving, I will agree with you – but not for the reason you might think.

I want to put to you that it *is* moving – *through* time.

This concept – moving through time – appears to be at odds with the standard model. The standard model will say two things. It will say that time passes. And it will say that time is bound up in the fabric of spacetime. OK, I exaggerate! I have heard the expression, 'travelling *through* time', used when describing the standard model. But it seems to depend on which part of the standard model is being talked about.

But it's ok.

The TRADD model casts time as a rescaled w-direction.

It's a direction.

What if you only had access to two directions? If you only had a two-dimensional world in x and y, then it wouldn't matter in which direction you moved – you would move *through* it. You would move *through* the x-y plane. You would move *through* your own two-dimensional world. If you only moved in the x-direction in the plane, you could say that you were moving through x. If you only moved in the y-direction in the plane, you could say that you were moving through y. Hold that thought.

Clearly, we have access to three directions. When we move around in our three-dimensional world cast in the x, y, and z directions, we say we move *through* it. We move through a volume. It doesn't matter which direction we move in – we are always moving through a volume. Our three-dimensional world is a volume and when we move, we move *through* it. Even if you are unable to access the up or down z-direction, and you're only moving around in the x-y plane, you're still moving through the volume. *Through it!* And if you were moving only in the x-direction, you're still moving through the volume. And you could say you're moving through x.

Are you getting the idea?

Our new TRADD model adds a fourth direction. The w-direction. Together with our x, y, and z directions, the w-direction makes up a four-dimensional world – a hyperspace. Three dimensions make a volume. Four dimensions make a hyperspace. Again, it doesn't matter in which direction we move. Whenever we move, we are moving *through* the hyperspace. We are moving through our four-dimensional world. And if we move in the w-direction, we could say we're moving through w.

We can call the hyperspace, spacetime if you like. It won't be the same spacetime as in the standard model, but it's a good proxy. So, we move through spacetime whenever we move through space. That we cannot access the w-direction is of no consequence. Even if restricted to the volume of x, y, and z, whenever we move through that volume, we are still also moving through hyperspace – *through* spacetime.

Of course, if we are not moving through space – then we are standing still, in space. But we will never know if we are standing still in space or not. That's because we measure the movement of ourselves or objects, relative to something else. You move along in your car relative to the road. If you stand still, you have no sense of the surface of the earth spinning on its axis and you moving around with it. And out in deep space, if your spaceship is following another one at a constant separating distance, then relative to each other, neither are moving.

But there's a problem. According to our TRADD model, our entire three-dimensional world is moving in the w-direction. And, as we have already said, whichever direction we move in, we would be moving *through* hyperspace. So, if we are already moving in the w-direction, then we are already moving *through* hyperspace. And we could say we were moving through w. So what's our speed? How fast are we moving in the w-direction. Well, according to the TRADD model, we always move at a rate of 1 light-year per year.

One light-year per year. That doesn't sound right, does it? It's awfully fast. It's light-speed. So how does that work? Well, we rescaled the w-direction, so it no longer counts off distance in light-years, but time in years. And recall we said it doesn't matter what units you put on the axis, nor what scale you use – it doesn't change what the axis is. So, if it's a direction, it will always be a direction, no matter how it's scales or what units are used. Now here's the concept that we really need to get our heads around. For every year that we move upward along the time-axis, that equates to one light-year in distance. Put another way, if we could put two scales on the time-axis, one in time units and another in distance units, then we will see that 1 year in time units is equivalent to 1 light-year in distance units. By the time we have moved up the time-axis for a period of 1 year, we will have travelled a *distance* of 1 light-year in the w-direction – in the time-direction. And no matter that our model talks about a year elapsing at different rates for different locations in space. Once a year has elapsed, we will have travelled 1 light-year in the w-direction. That's in the time direction. One light-year per year ... *through* time.

Well, almost. You see, there's one last detail to explore and explain. When we move through a *volume*, we almost always move through the z-direction – the up and down direction – either a lot or a little bit. It is only when our movement is restricted to an x-y plane in the volume, that we don't *also* move in the z-direction. Not even a little bit. So, when we move through a *hyperspace*, we similarly almost always move through the w-direction – either a lot or a little bit. It is only when our movement is restricted to either an x-y-z volume or an x-y, y-z, or x-z plane, that we don't also move in the w-direction, not even a little bit. But because the TRADD model says that our x-y-z volume is moving in the w-direction, then we are indeed moving through the w-direction – *through* time – all the time.

OK. So if we are always moving through time and always at a speed of 1 light-year per year, then what does that mean and why is it an important consideration?

It means that in the context of spacetime, we are moving *through* time, all the time, at a speed of 1 light-year per year – at light-speed, even as we're standing still in our three-dimensional volume – standing still in space.

This is important for when we want to move in any of the x, y, or z directions. For when we want to travel through our volume – through space.

Imagine you are travelling north along a highway at 100 km per hour and you come upon an intersection – a crossroad – and you want to turn right. You want to travel east. Well, you can't just keep on driving at 100 km per hour and then suddenly turn right at the intersection and be travelling east at 100 km per hour. You first need to slow down, almost to a stop, turn right, and then speed up again. Do you see the problem? If you're already travelling in the w-direction at 1 light-year per year and you want to turn right, in the x-direction, you also need to slow down, make the turn, and then speed up again. Do you see the problem now? I bet you do!

The problem is – we *can't* slow down! We can't access the w-direction. Although we are moving in the w-direction, we can't *travel* in the w-direction. We move in the w-direction because we are *being* moved. Not because we are moving ourselves. We have no control. Despite an interpretation here or there in the standard model, we cannot slow our speed through time. We cannot stop time. Nor can we go faster through time. We can only move in various combinations of the directions x, y, and z. We can only move ourselves though the volume we live in. We can only navigate space – not time.

So what happens when we move off in the x-direction? Well, we change our direction through spacetime. Let me explain. We are already moving in the w-direction at light-speed. To travel in the x-direction, we can't first stop travelling in the w-direction. We can't stop moving through time. We do not have access to the w-direction. But when we move off in the x-direction, we do that by steering ourselves in the x-direction, even as we are moving in the w-direction. We continue travelling at light-speed. But we change our direction more toward the x-direction in spacetime and consequently less toward the w-direction having our time-axis. Doing that means it will take longer for a year to elapse for us against the time-axis than if we stood still. For if we didn't move. The following Charts show what happens when we increase our velocity through space.

It's important to note – 'increase our velocity through space'. We are not talking about increasing our velocity through spacetime. We cannot move our selves through spacetime. We can only move ourselves through space. Through our three-dimensional world. Through our volume made up of directions x, y, and z. And, as you will see, we move our selves through space, by changing our direction through spacetime.

When our velocity through space is zero

If we take a position in spacetime and pick up our motion or lack of it through spacetime at t=0 then we can see what happens, for example, if we don't move. If we stand still. We progress upward along the time-axis for 1 year, after which we will have travelled a 'distance' of 1 light-year. Remember, our time-axis is actually a rescaled spatial axis in the w-direction, along which 1 year in the time direction is equivalent to 1 light-year in the w-direction.

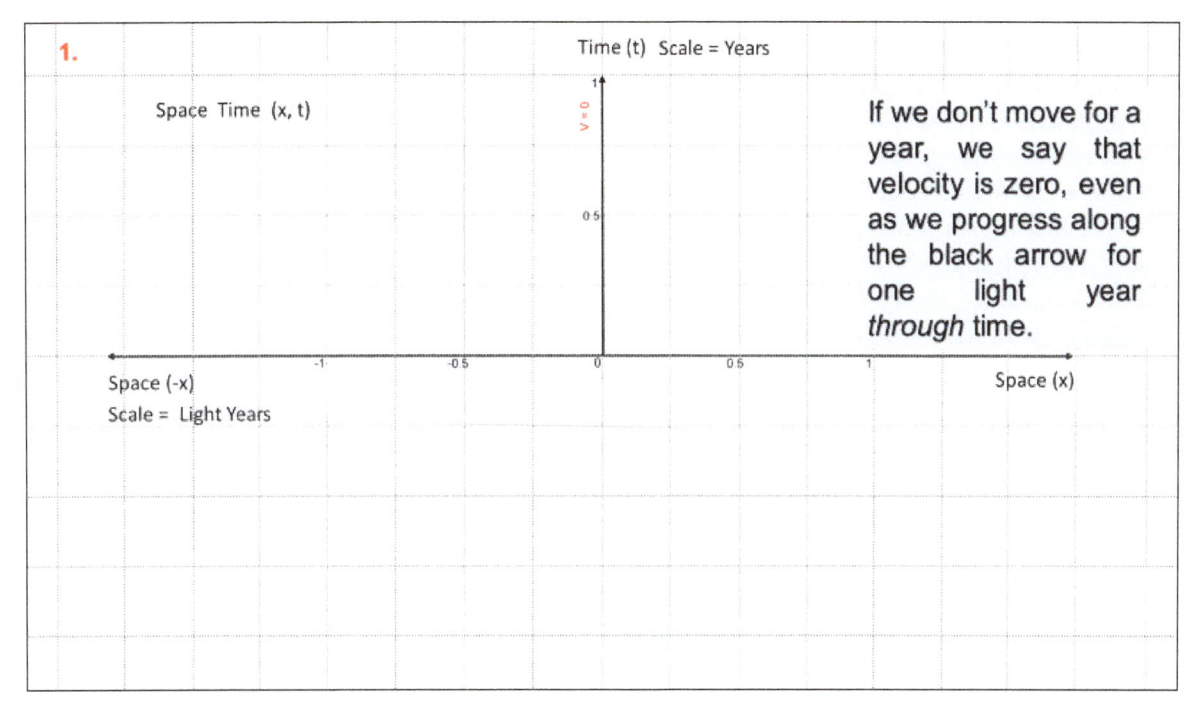

Figure 89.

If we were able to measure the *distance* travelled in the w-direction, we would measure 1 light-year after we have moved up along the time-axis for 1 year. Time is a direction through spacetime, as are x, y, and z. However, x, y and z, are directions through space, whereas time is not. Put a different way, x, y and z, are directions through space, whereas w is not. You see, x, y, and z, make up the directions in space and x, y, z, and w, make up the directions in spacetime. We could have called it space-w, but because we perceive movement in the w-direction as time, we call it spacetime.

When our velocity through space is 0.25 light-speed

If before t=0 we wanted to accelerate to 0.25 light-speed in the x-direction, the way we do that is to push off in the x-direction. We are already moving through time at light-speed. So the best we can do by pushing in the x-direction is to change our direction through spacetime. In the chart below, we changed our direction through spacetime so that we reached our desired velocity of 0.25 light-speed in the x-direction just as we pass through t=0. Thereafter, still travelling at light-speed, after travelling for 1 light-year through spacetime, we will have reached a distance along the x-axis, through space, of 0.25 light-years. It's challenging, but if you think about it, you will see how that works.

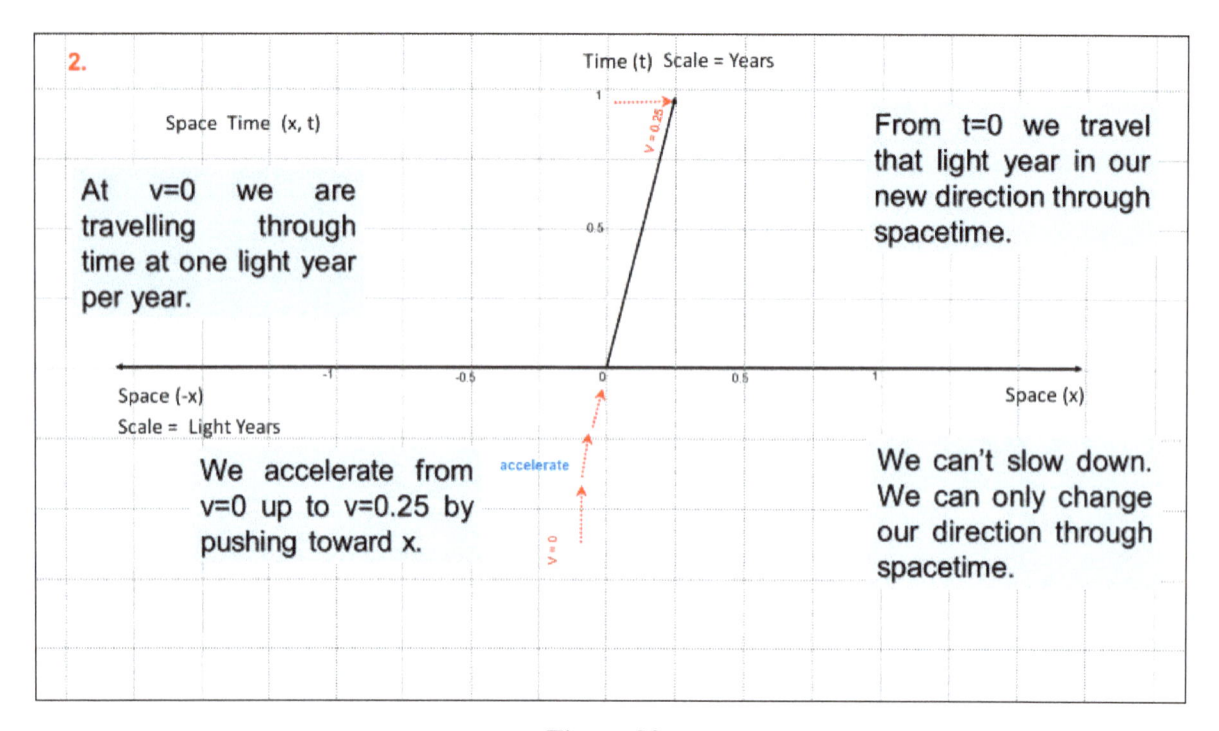

Figure 90.

First, think about this. If we didn't push off in the x-direction we would have travelled 1 year from t=0 to t=1, in the time-direction, which before renaming it was the w-direction, and which before rescaling it, had units of distance. That is, in 1 year, we would have travelled 1 light-year in the w-direction.

We perceive travel in the w-direction as time. So, in that upward direction, there is a distance travelled (1 light-year) as well as a duration of travel (1 year). Hence our speed through spacetime of 1 light-year per year.

But we did push off. So instead of that 1 light-year being travelled in the w-direction, through spacetime, (through time), it was travelled partly in the x-direction, through space. We can't change our speed through spacetime, but in pushing off we did change our speed through space, in the x-direction. Then, having travelled our 1 light-year through spacetime, we find we've travelled 0.25 light-years in the x-direction, through space, and less than 1 light-year in the time direction, through spacetime.

Think of a car travelling north at 100 km per hour. You don't change speed but steer it eastward until, per hour, it is travelling 25 km per hour eastward, but less than 100 km per hour northward.

When our velocity through space is 0.5 light-speed

I'm going to repeat myself here, as for what I wrote, when our velocity is 0.25 light-speed. Bear with me. Repeating myself is deliberate, and important, to help you understand what's going on.

If before t=0 we wanted to accelerate to 0.5 light-speed in the x-direction, the way we do that is to push off in the x-direction. We are already moving through time at light-speed. So the best we can do by pushing in the x-direction is to change our direction through spacetime. In the chart below we changed our direction through spacetime so that we reached our desired velocity of 0.5 light-speed in the x-direction just as we pass through t=0. Thereafter, still travelling at light-speed, after travelling for 1 light-year we will have reached a distance along the x-axis of 0.5 light-years.

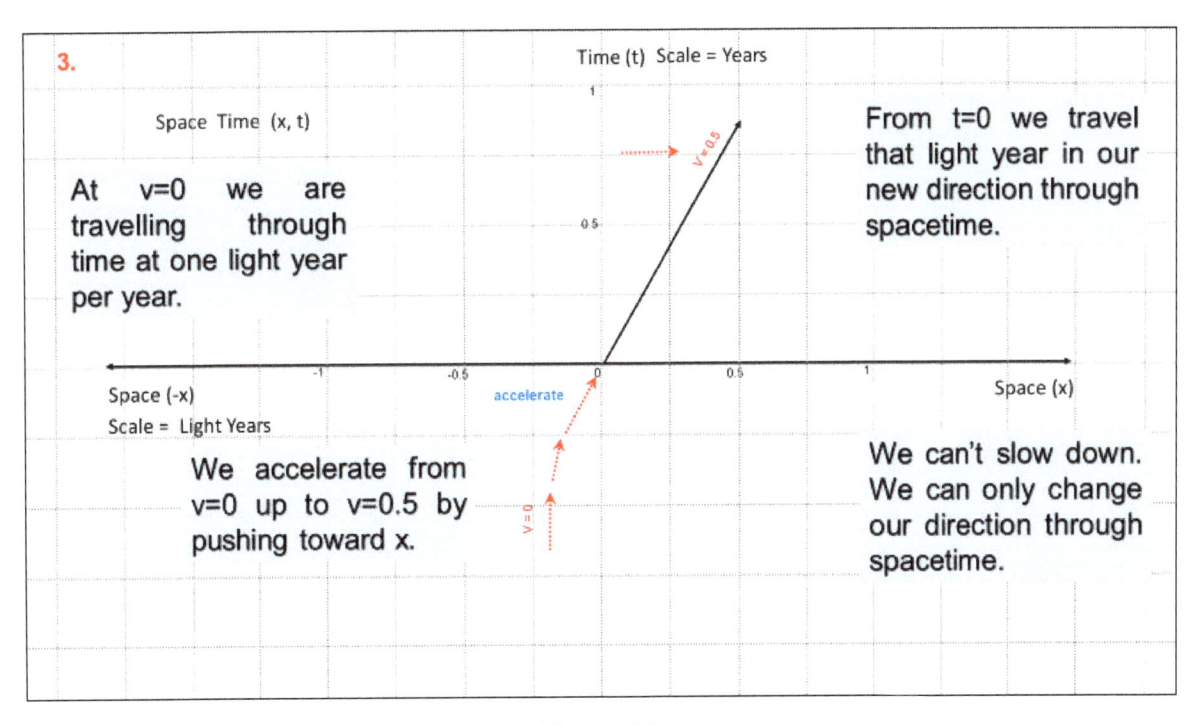

Figure 91.

When our velocity through space is 0.75 light-speed

If before t=0 we wanted to accelerate to 0.75 light-speed in the x-direction, the way we do that is to push off in the x-direction. We are already moving through time at light-speed. So the best we can do by pushing in the x-direction is to change our direction through spacetime. In the chart below we changed our direction through spacetime so that we reached our desired velocity of 0.75 light-speed in the x-direction just as we pass through t=0. Thereafter, still travelling at light-speed, after travelling for 1 light-year we will have reached a distance along the x-axis of 0.75 light-years.

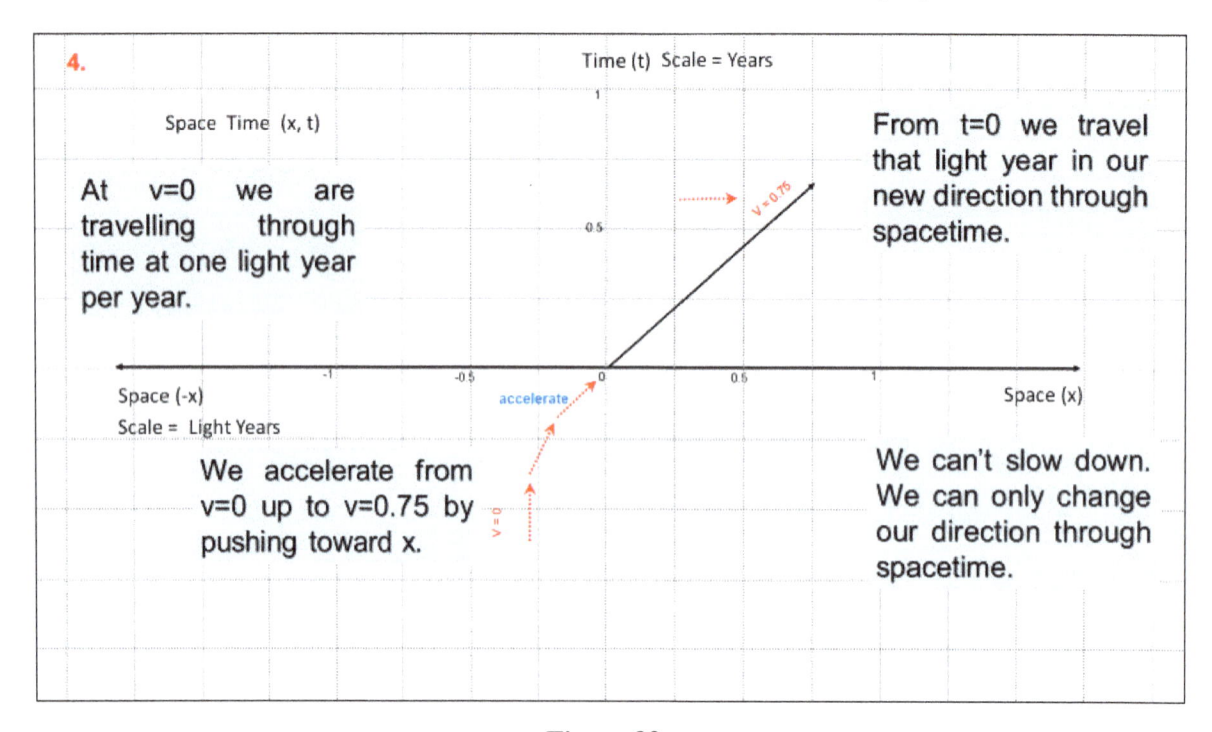

Figure 92.

What is time?

Our speed through spacetime is constant

Our speed through spacetime never changes. It is constant. When our velocity through space is zero, we travel 1 light-year in 1 year – but in the time direction – through spacetime. But when our velocity through space is greater than zero, say, 0.5 light-speed, clearly our velocity through time is less than light-speed. Some of our speed is directed in the x-direction. But here's the thing. Whilst our velocity through time reduces as our velocity through space increases, our *speed* through spacetime doesn't change. It remains constant.

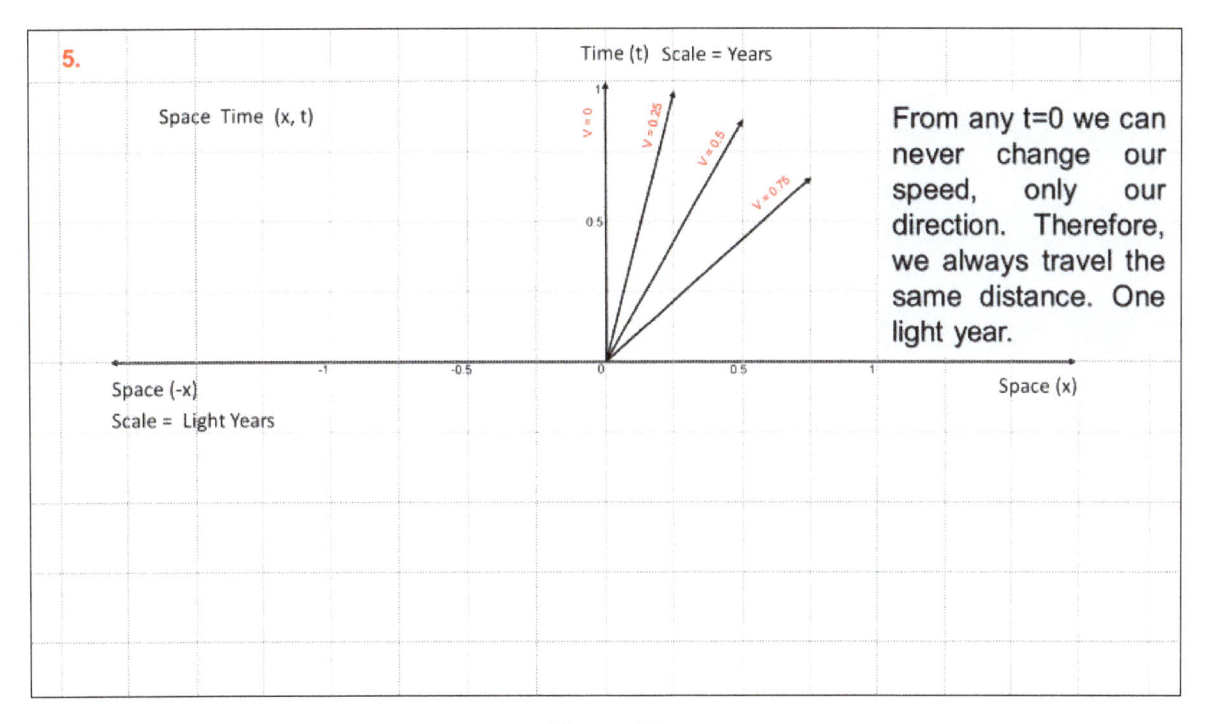

Figure 93.

To understand how this works, we can look at the chart above. All journeys pass through x=0 and t=0 at the same point in space and time. Thereafter, each journey completes 1 light-year of travel. In the chart, each journey travels at a different velocity through space, and at a different velocity through time. But the progress outward from x=0 and t=0 is uniform. By that, I mean that when the journey of v=0 has progressed half a light-year, the journey of v=0.25 will have progressed half a light-year too. No more. No less. Each of the journeys will have progressed by the same distance at any instant in spacetime. So, through spacetime, whilst their velocities through time and through space may differ, their speed through spacetime relative to one another always remains the same.

We travel at the speed of space

If we look at the chart below, we see that the destinations of journeys after travelling 1 light-year, are located on the circumference of a circle. The circumference represents the radial displacement of expanding spacetime, from x=0 and t=0, after 1 year. Take a point in spacetime and expand out from that point. After 1 light-year in the x-direction, it will have expanded 1 year in the time direction. It would seem then, that as spacetime expands, our travellers are always pushed outward, along with the expansion. Our speed, then, through spacetime, is the speed of space!

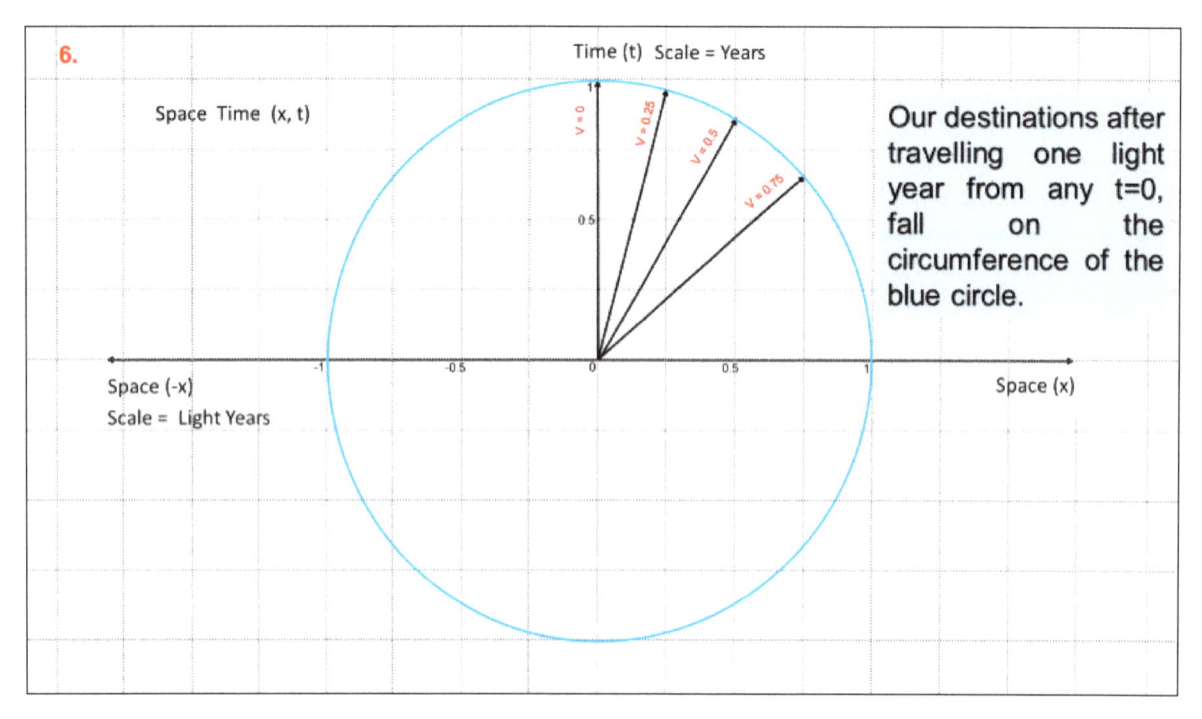

Figure 94.

The why and how of time dilation

If you're like me, you might be fascinated by the term 'time dilation'. Time slowing down. What's fascinating about it goes beyond the concept itself. We know it occurs in a gravitational field *and* when we move. But science gives different reasons for each of these occurrences. We've been able to measure it. It's a physical phenomenon. Yet, the very idea of time slowing down is so alien to our notions of anything that might make sense, that what's fascinated me more is the ease with which we accept these ideas when we don't even know what time is.

Earlier, we dealt with time dilation in a gravitational field. Here, we are dealing with time dilation that occurs when we move. There are some explanations out there and demonstrations showing why and how time dilation works, but these have more to do with measuring how light travels distances between moving objects. Whilst these, like magic, appear to satisfy our curiosity, in the context of science unable to even agree what time is, one wonders why we so easily accept them.

In our TRADD model, 'what time is' *has* been defined. We now know what time *is*. Therefore, the model ought to be able to shed some light on the 'why' and 'how' of time dilation. And it should accord with the measurements presented in the standard model. There should be no contradiction. But there should be an explanation. And as it happens, there is.

In the chart below, we can see that the light-year travelled by each of the journeys terminates at a destination on the circumference of a circle with a radius of one. Having travelled through spacetime, the destinations have different space coordinates and different time coordinates. The values of the time coordinates can be read off the time-axis and the values of the space coordinates can be read off the space-axis. We can see now that steering through spacetime to obtain movement through space, trades some of the movement in the time-direction for movement in the x-direction. The difference along the time-axis between the time taken by a journey and 1 year, represents the value of time-dilation referred to in the standard model.

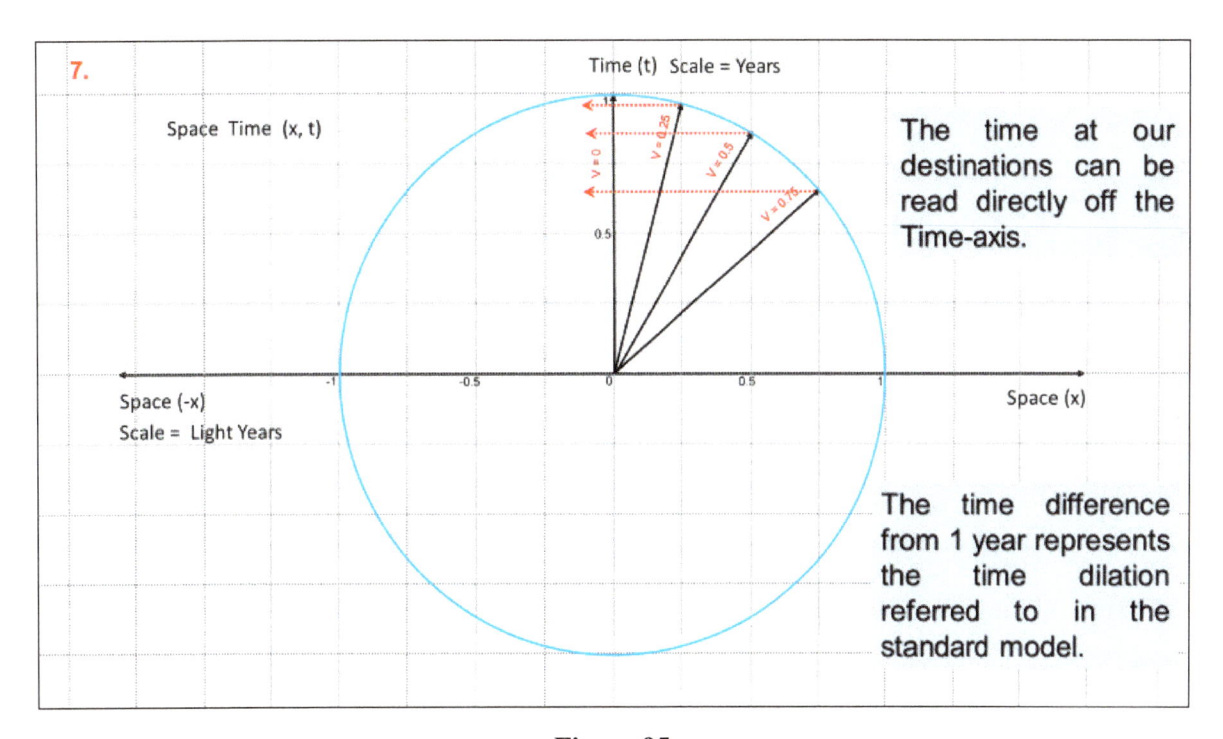

Figure 95.

Calculating time dilation in the TRADD model

Whilst we could simply use the chart to read off what the time is on the travelled spaceship after 1 year, if we recognize the triangle in the circle, we can do the calculation instead. We want to find the value on the time-axis at the time the journey travelling at 0.75 light-years per year had progressed for 'one year'. I put 'one year' in double quotes, to remind us that velocity through space is a relative. The journey we are looking at is relative to the journey taken at v=0. After 1 year, those at the destination after the v=0 journey are interested to know how much time has elapsed for those at their destination after the v=0.75 journey.

We can calculate this by noting the top edge of the triangle is 0.75 and the long edge of the triangle is 1. We want to find the length of the left edge. It is simply the positive square root of 1 squared minus 0.75 squared, which comes out to the positive square root of 0.4375, or around 0.6614 years.

This is the same value given by the standard model. There is no contradiction. What is new is our perspective. We can see that travelling through spacetime as we have, has set us on a path that diverges from the path taken if we did not move. When we diverge due to motion, we move apart in both the time and the space directions. Because we move, we move away in space, and we move away in time. But we note, we always move away in the negative time-direction. We move fastest through time when we don't move through space. Once we move through space, then relative to not moving, time slows down. Or as I like to put it, when we move, we spend less time than when we don't.

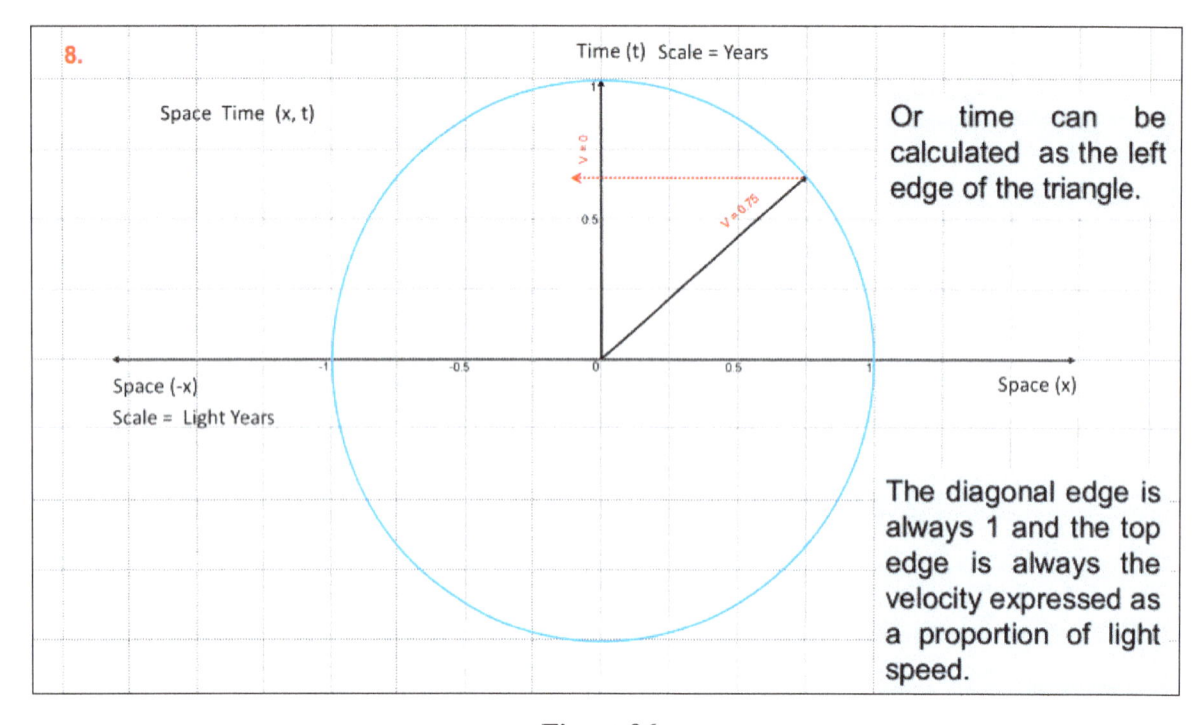

Figure 96.

The standard model uses some math, the Lorentz transformation, to calculate time dilation. The Lorentz transformation formula can be extracted directly from our chart, but I wanted to avoid using math. So suffice to say, allowing 1 year of travel at v=0 and expressing velocity as a proportion of 1 light-year, is key if comparing TRADD time dilation to the time dilation in the standard model. You can with a little effort, extract the Lorentz transformation from the graph.

Final thoughts

There's a certain anti-climactic relief that washes over me as I look back over the material presented in this book. While it's nice to think that the work's been done, there's always the lingering worry about how well I've pitched the information to make it accessible. And how well I've managed to make the subject interesting enough to hold your attention. My initial goal was to get the ideas that were swimming around in my head for years, out of my head, and be done with it. A secondary goal was to have these ideas documented so those close to me could get to those insights which disrupted my otherwise dependable demeanour when they emerged. And finally, I wanted to present these ideas in a way that most people would be able to understand them. I have a saying: 'There's no such thing as a dumb human being.' But sometimes we feel inadequate comprehending something. And in my opinion, most of the time, it's because of the way the information is presented – not our ability to understand.

And then there's what I have not written. What I've left out. And there's quite a lot. Where there were opportunities to explain things in more detail or to expand or extend an idea, or introduce a new but related idea, I have variously resisted. Partly due to the extra work needed to crystalise these to make them relatable, and partly to stay within the lines guiding the content of the book. You've all probably encountered people who once they get started on their pet topic, will talk for ever and never shut up. Well, that's me. But it can't be me if I'm asking you to read about it. At some point – enough is enough.

So it's time to gather thoughts and summarise.

Total relativity and dimensional dynamics (TRADD) is a cosmological model incorporating space, gravity, light, and time. By offering different perspectives on what we already know, the model attempts to make sense of counter intuitive phenomena by suggesting *why* and *how* things happen as they do. I set out to understand what time was, but strayed to looking at space, gravity, and light. Our universe is expanding, gravity bends space and slows time, light travels at a constant speed, and time … well … it just passes – doesn't it?

Space

If we endow space with attributes, it could then behave according to a simple set of rules. We could refer to the way space follows these rules as dimensional dynamics. We could then model space using the concepts of dimensional dynamics. So that's what we did. Dimensional dynamics allows the way space expands to be different from the 'Big Bang' concept which requires a beginning called a singularity as a starting point. Representing space as *something* rather than *nothing* provides a natural medium for the propagation of gravity waves and light. It sets the stage as it were, for new descriptions of gravity, light, and time.

Considering space exclusive of time and *why* and *how* space expands was central in the pursuit of a new description for it.

Gravity

The description of gravity under general relativity did not map onto the TRADD model because the bending of spacetime connotes a bending of time which the model does not support. A distaste for 'action at a distance' further motivated a rejection of the description under general relativity. Using dimensional dynamics, we found a unique way to describe gravity that would map onto the TRADD model whilst retaining its measurable effects and give relevance to 'action at a distance'. This new description of gravity offers different insights into what black holes might be and *why* and *how* they form. A mass falling through time at greater than one year per year (where a year is according to the observer's local time) is an environment from which light travelling at one light-year per year, can never reach the observer, simply because each year, there is more than one additional year that light must travel to complete the journey.

Considering 'action at a distance' and *why* and *how* we feel gravity as a force was central in the pursuit of a new description for it.

Light

The idea that the approach speed of light when measured is always the same value is counter intuitive. That its speed somehow sets a speed limit is even more bizarre. That we can measure it as a wave and as a particle suggests it *travels* as a wave *and* as a particle. We know that waves and particles progress differently. The idea that both ways of progressing occur concurrently makes no sense at all. A leaf floating down a river progresses differently to a wave rippling outward from where a thrown pebble entered the water.

We take measurements and seem happily confounded by the results, so long as we can use them to make reliable predictions. Using the model, and with a little imagination and persistence, we found a new way to describe how light might move. The constant approach speed of light can finally make sense – it's no longer counter intuitive. The model suggests light travels like a wave by hitching a ride on the expanding space. It is expanding space that spreads out, propagating like a wave. It is the *constant rate* at which space expands that sets a speed limit. Light, travelling along with the expansion, can only travel at *that* speed.

A light wave collapses when detected, transferring its energy. We call the energy carried by a light wave, a photon. Photons are what we detect as particles. If we were able to animate the progress of light until detected, the model would show light travelling through space and time, but always behind us in time. We never 'see' light travel through space even as that is our perception. Light always approaches our position in space and time from the 'past'.

That, too, is a counter-intuitive concept until we have a different description of time. Hence, until detected, light travels as a wave always arriving as a particle imparting energy. When you think about how the double slit experiment works, it just may start to make sense. Through the slits as a wave, waves on the other side of the slits interfere with one another, then finally from the past into the present detected as energy – a packet of energy – a photon – a particle. For light, the *why* and *how* its approach speed is always the same speed was central in the pursuit of a new description for it.

Time

If we looked west and saw a field, then travelled west, in respect of the field, we would move through it. The TRADD model developed a four-dimensional construct with axes w, x, y, and z. We are all familiar with the three directions x, y, and z. They describe the three-dimensional volume that *is* the space that we live in and experience. When we look out at the universe, we see a vast volume of space. And if we travelled in any direction, we would move through it.

The TRADD model suggests there is another direction – the w-direction. It suggests the metaphor, that if we were able to look in the w-direction we would see time, and if we travelled in the w-direction, then in respect of time, we would move through it. Of course we can't look in the w-direction. We can't see time. But the model does say we travel in the w-direction - involuntarily. In respect of time then, we do move through it.

Our notions of past, present, and future stem from our recognition of change. The model couples change to the expansion of space, and couples the expansion of space to its movement through time. Any description of time ought to account for our measurements of it. The model suggests *why* and *how* time slows down due to our closer proximity to mass (general relativity) or when we speed up (special relativity).

Considering time exclusive of space and *why* and *how* we measure and experience time, was central in the pursuit of a new description for it.

TRADD

If we accept our new model with its different descriptions of what we already know and measure. If we trust that spatial dimensions exist in an inherent state of tension with one another. If tension implies force. If force implies the presence of energy. If we are confident that mass and energy are equivalent. Then perhaps space possesses energy to manifest dimensional tensions and inter-dimensional tensions. Space would need to release that energy when no longer needed. When no longer participating in the forces balancing and rebalancing these tensions. Space itself, where there are points of imbalance in its dimensional dynamics, must give up its energy. Energy, at points of extreme imbalance, might condense into mass. These are merely suggestions based on attributes embedded in the model. These attributes embody the mechanics of spacetime. Space, then, could be nothing, or something. Or more than that – *space* could be everything.

Notes

1. *QED, The Strange Theory of Light and Matter*: Richard Feynman, 1985, Princeton University Press
   a. Chapter 2, 'Photons: Particles of Light', page 37
      i. (Quantum electrodynamics 'resolves' this wave-particle duality by saying that light is made of particles (as Newton originally thought), but the price of this great advancement of science is a retreat by physics to the position of being able to calculate only the *probability* that a photon will hit a detector, without offering a good model of how it actually happens.)
2. *Gravitation*: Charles W. Misner, Kip S. Thorne, John Archibald Wheeler, 2017, Princeton University Press
   a. '*Mass acts on spacetime, telling it how* to curve. Spacetime acts on mass, telling it how to move.'

Acknowledgments

Burr Dodd: Burr mentored me through the years, steering me in my pursuit to understand the nature of time. I thank him for his patience, encouragement, and guidance. Thanks Burr. It was Burr's idea that the book needed a subtitle to better explain its contents.

Lolo Houbein: Lolo, my mother, herself an author, always listened to my latest ideas about space and time. I don't believe she or anyone else for that matter, really understood what I was talking about. But she always believed that I believed what I was talking about and that helped me believe it too. Thanks Ma.

Everyone Else: The list of friends and workmates who suffered my enthusiasm for the material in this book is long and was accumulated over 40 years. I feel fortunate so many individuals tolerated my lengthy explanations for the most abstract considerations. They were generous with their time and with their comments. They know who they are. Special mentions to the following who were kind enough to comment on early drafts of this book. I doubt anyone thought my ideas were in anyway valid in a scientific sense – but they thought them interesting enough to support my progressing with them: Mark Rance, Fred Accary, Taka Muraoka, Alex Paykin, Ean Sugarman, Victor Segal, John Strumilla, Paul Ng.

Cover design: Ashley Menezes

Cover Blurb: Lodie Webster